T0257874

Heat Transfer Sourcebook

Heat Transfer Sourcebook

Edited by **Nathan Rice**

LANRYE
INTERNATIONAL

New Jersey

Published by Clanrye International,
55 Van Reypen Street,
Jersey City, NJ 07306, USA
www.clanryeinternational.com

Heat Transfer Sourcebook
Edited by Nathan Rice

International Standard Book Number: 978-1-63240-297-4 (Hardback)

Contents

Preface

This book includes an extensive examination of heat transfer phenomenon and its potential. In today's world, unprecedented industrial growth, rapid urbanization, developments in transportation and persistent human habits have resulted in acute energy shortage. In such a scenario, energy saving can be vitally affected by efficient transfer of heat. Effective heat exchange tools have a great impact on industries, household requirements, offices and transportation. Taking these factors into account, this book has incorporated a structured study of the general aspects of heat transfer phenomena, convection, boiling and condensation by various means.

The information contained in this book is the result of intensive hard work done by researchers in this field. All due efforts have been made to make this book serve as a complete guiding source for students and researchers. The topics in this book have been comprehensively explained to help readers understand the growing trends in the field.

I would like to thank the entire group of writers who made sincere efforts in this book and my family who supported me in my efforts of working on this book. I take this opportunity to thank all those who have been a guiding force throughout my life.

Editor

General Aspects of Heat Transfer

Measurements of Local Heat Flux and Water-Side Heat Transfer Coefficient in Water Wall Tubes

Jan Taler and Dawid Taler

Additional information is available at the end of the chapter

1. Introduction

Measurements of heat flux and heat transfer coefficient are subject of many current studies. A proper understanding of combustion and heat transfer in furnaces and heat exchange on the water-steam side in water walls requires accurate measurement of heat flux which is absorbed by membrane furnace walls. There are three broad categories of heat flux measurements of the boiler water-walls: (1) portable heat flux meters inserted in inspection ports [1], (2) Gardon type heat flux meters welded to the sections of the boiler tubes [1-4], (3) tubular type instruments placed between two adjacent boiler tubes [5-14]. Tubular type and Gardon meters strategically placed on the furnace tube wall can be a valuable boiler diagnostic device for monitoring of slag deposition. If a heat flux instrument is to measure the absorbed heat flux correctly, it must resemble the boiler tube as closely as possible so far as radiant heat exchange with the flame and surrounding surfaces is concerned. Two main factors in this respect are the emissivity and the temperature of the absorbing surface, but since the instrument will almost always be coated with ash, it is generally the properties of the ash and not the instrument that dominate the situation. Unfortunately, the thermal conductivity can vary widely. Therefore, accurate measurements will only be performed if the deposit on the meter is representative of that on the surrounding tubes. The tubular type instruments known also as flux-tubes meet this requirement. In these devices the measured boiler tube wall temperatures are used for the evaluation of the heat flux q_m. The measuring tube is fitted with two thermocouples in holes of known radial spacing r_1 and r_2. The thermocouples are led away to the junction box where they are connected differentially to give a flux related electromotive force.

The use of the one dimensional heat conduction equation for determining temperature distribution in the tube wall leads to the simple formula

$$q_m = \frac{k\left(f_1 - f_2\right)}{r_o \ln\left(r_1/r_2\right)}. \tag{1}$$

The accuracy of this equation is very low because of the circumferential heat conduction in the tube wall.

However, the measurement of the heat flux absorbed by water-walls with satisfactory accuracy is a challenging task. Considerable work has been done in recent years in this field. Previous attempts to accurately measure the local heat flux to membrane water walls in steam boilers failed due to calculation of inside heat transfer coefficients. The heat flux can only be determined accurately if the inside heat transfer coefficient is measured experimentally.

New numerical methods for determining the heat flux in boiler furnaces, based on experimentally acquired interior flux-tube temperatures, will be presented. The tubular type instruments have been designed to provide a very accurate measurement of absorbed heat flux q_m, inside heat transfer coefficient h_{in}, and water steam temperature T_f.

Two different tubular type instruments (flux tubes) were developed to identify boundary conditions in water wall tubes of steam boilers.

The first meter is constructed from a short length of eccentric bare tube containing four thermocouples on the fire side below the inner and outer surfaces of the tube. The fifth thermocouple is located at the rear of the tube on the casing side of the water wall tube. First, formulas for the view factor defining the heat flux distribution at the outer surface of the flux tube were derived. The exact analytical expressions for the view factor compare very well with approximate methods for determining view factor which are used by the ANSYS software. The meter is constructed from a short length of eccentric tube containing four thermocouples on the fireside below the inner and outer surfaces of the tube. The fifth thermocouple is located at the rear of the tube (on the casing side of the water-wall tube). The boundary conditions on the outer and inner surfaces of the water flux-tube must then be determined from temperature measurements at the interior locations. Four K-type sheathed thermocouples, 1 mm in diameter, are inserted into holes, which are parallel to the tube axis. The thermal conduction effect at the hot junction is minimized because the thermocouples pass through isothermal holes. The thermocouples are brought to the rear of the tube in the slot machined in the tube wall. An austenitic cover plate with the thickness of 3 mm – welded to the tube – is used to protect the thermocouples from the incident flame radiation. A K-type sheathed thermocouple with a pad is used to measure the temperature at the rear of the flux-tube. This temperature is almost the same as the water-steam temperature.

The non-linear least squares problem was solved numerically using the Levenberg–Marquardt method. The temperature distribution at the cross section of the flux tube was determined at every iteration step using the method of separation of variables.The heat transfer conditions in adjacent boiler tubes have no impact on the temperature distribution in the flux tubes.

The second flux tube has two longitudinal fins which are welded to the eccentric bare tube. In contrast to existing devices, in the developed flux-tube fins are not welded to adjacent water-wall tubes. Temperature distribution in the flux-tube is symmetric and not disturbed by different temperature fields in neighboring tubes. The temperature dependent thermal conductivity of the flux-tube material was assumed. An inverse problem of heat conduction was solved using the least squares method. Three unknown parameters were estimated using the Levenberg-Marquardt method. At every iteration step, the temperature distribution over the cross-section of the heat flux meter was computed using the ANSYS CFX software. Test calculations were carried out to assess accuracy of the presented method. The uncertainty in determined parameters was calculated using the variance propagation rule by Gauss. The presented method is appropriate for membrane water-walls.

The developed meters have one particular advantage over the existing flux tubes to date. The temperature distribution in the flux tube is not affected by the water wall tubes, since the flux tube is not connected to adjacent waterwall tubes with metal bars, referred to as membrane or webs. To determine the unknown parameters only the temperature distribution at the cross section of the flux tube must be analysed.

2. Tubular type heat flux meter made of a bare tube

Heat flux meters are used for monitoring local waterwall slagging in coal and biomass fired steam boilers [5-19].

The tubular type instruments (flux tubes) [10-14,19] and other measuring devices [15-18] were developed to identify boundary conditions in water wall tubes of steam boilers. The meter is constructed from a short length of eccentric tube containing four thermocouples on the fire side below the inner and outer surfaces of the tube. The fifth thermocouple is located at the rear of the tube on the casing side of the water wall tube.

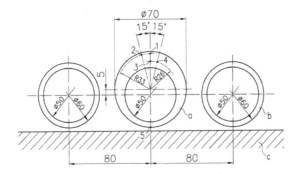

Figure 1. The heat flux tube placed between two water wall tubes, a – flux tube, b – water wall tube, c – thermal insulation

The boundary conditions at the outer and inner surfaces of the water flux-tube must then be determined from temperature measurements at the interior locations. Four K-type sheathed thermocouples, 1 mm in diameter, are inserted into holes, which are parallel to the tube axis. The thermal conduction effect at the hot junction is minimized because the thermocouples pass through isothermal holes. The thermocouples are brought to the rear of the tube in the slot machined in the protecting pad. An austenitic cover plate with the thickness of 3 mm welded to the tube is used to protect the thermocouples from the incident flame radiation. A K-type sheathed thermocouple with a pad is used to measure the temperature at the rear of the flux-tube. This temperature is almost the same as the water-steam temperature. A method for determining fireside heat flux, heat transfer coefficient on the inner surface and temperature of water-steam mixture in water-wall tubes is developed. The unknown parameters are estimated based on the temperature measurements at a few internal locations from the solution of the inverse heat conduction problem. The non-linear least squares problem is solved numerically using the Levenberg–Marquardt method. The diameter of the measuring tube can be larger than the water-wall tube diameter. The view factor defining the distribution of the heat flux on the measuring tube circumference is determined using exact analytical formulas and compared with the results obtained numerically using ANSYS software. The method developed can also be used for an assessment of scale deposition on the inner surfaces of the water wall tubes or slagging on the fire side. The presented method is suitable for water walls made of bare tubes as well as for membrane water walls. The heat transfer conditions in adjacent boiler tubes have no impact on the temperature distribution in the flux tubes.

2.1. View factor for radiation heat transfer between heat flux tube and flame

The heat flux distribution in the flux tube depends heavily on the heat flux distribution on its outer surface. To determine the heat flux distribution q as a function of angular coordinate ϕ, the analytical formulas for the view factor ψ, defining radiation interchange between an infinitesimal surface on the outer flux tube circumference and the infinite flame or boiler surface, will be derived. The heat flux absorbed by the outer surface of the heat flux tube $q(\varphi)$ is given by

$$q(\varphi) = q_m \psi(\varphi).$$

(2)

The specific thermal load of the water wall q_m is defined as the ratio of the heat transfer rate absorbed by the water wall to the projected surface area of the water wall. The view factor is the fraction of the radiation leaving the surface element located on the flux tube surface that arrives at the flame surface. The view factor can be computed from

$$\psi = \frac{1}{2}(\sin \delta_1 + \sin \delta_2).$$

(3)

The angles δ_1 and δ_2 are formed by the normal to the flux tube at φ and the tangents to the flux tube and adjacent water-wall tube (Figures 2,4,6). Positive values of δ_1 are measured clockwise with respect to the normal while positive values of δ_2 are measured counterclockwise with respect to the normal. The radial coordinate r_o of the flux tube outer surface measured from the center 0 (Figure 2) is

$$r_o = e\cos\varphi + \sqrt{b^2 - e^2(\sin\varphi)^2}. \tag{4}$$

where: e – eccentric (Figure 2), b – outer radius of flux-tube.

The angle φ_1 can be expressed in terms of the angle φ, flux tube outer radius b, and eccentric e (Figure 2)

$$\varphi_1 = \arcsin\left[\frac{\left(e\cos\varphi + \sqrt{b^2 - (e\sin\varphi)^2}\right)\sin\varphi}{b}\right], \quad \varphi_1 \leq \frac{\pi}{2}, \tag{5}$$

$$\varphi_1 = \pi - \arcsin\left[\frac{\left(e\cos\varphi + \sqrt{b^2 - (e\sin\varphi)^2}\right)\sin\varphi}{b}\right], \quad \frac{\pi}{2} \leq \varphi_1 \leq \pi. \tag{6}$$

First, the view factor for the angle interval $0 \leq \varphi_1 \leq \varphi_{1,l1}$ was determined

$$\psi = \frac{1 + \cos\varphi_1}{2}, 0 \leq \varphi_1 \leq \varphi_{1,l1}. \tag{7}$$

The limit angle $\varphi_{1,l1}$ (Figures 2 and 3) is given by

$$\varphi_{1,l1} = \arccos\frac{c - e}{b}, \tag{8}$$

where c is the outer radius of the boiler tube.

Next the view factor in the angle interval $\varphi_{1,l1} \leq \varphi_1 \leq \varphi_{1,l2}$ will be determined. The limit angle $\varphi_{1,l2}$ is: $\varphi_{1,l2} = \varphi_1(\varphi = \pi/2) = (\pi/2) + \arcsin(e/b)$ (Figure3). The view factor ψ is computed from Eq.(2), taking into account that (Figure 4)

$$\delta_1 = \frac{\pi}{2}, \quad \delta_2 = \frac{\pi}{2} - (\varepsilon + \varphi_1), \quad \varepsilon = \beta + \gamma - \frac{\pi}{2}, \quad x_i = b\sin\varphi_1, \quad x_i = b\cos\varphi_1,$$

$$\beta = \arcsin\frac{c}{\sqrt{(t - x_i)^2 + (y_i + e)^2}}, \quad \gamma = \arcsin\frac{t - x_i}{\sqrt{(t - x_i)^2 + (y_i + e)^2}}, \quad \varphi_{l1} \leq \varphi_1 \leq \varphi_{l2}, \tag{9}$$

where t is the pitch of the water wall tubes.

Next the view factor $\psi(\phi)$ is determined in the angle interval $\varphi_{1,l2} \leq \varphi_1 \leq \varphi_{1,l3}$ (Figures 3 and 5).

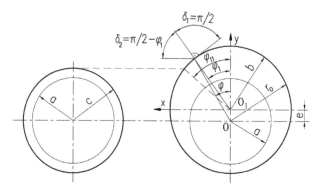

Figure 2. Determination of view factor in the angle interval $0 \leq \varphi_1 \leq \varphi_{1,l1}$

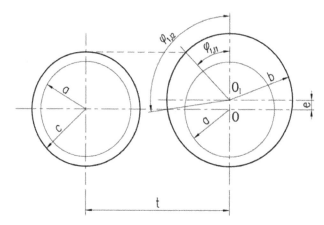

Figure 3. Limit angles $\varphi_{1,l1}$ and $\varphi_{1,l2}$

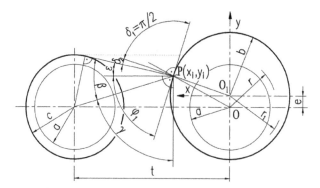

Figure 4. Determination of view factor in the angle interval $\varphi_{1,l1} \leq \varphi_1 \leq \varphi_{1,l2}$

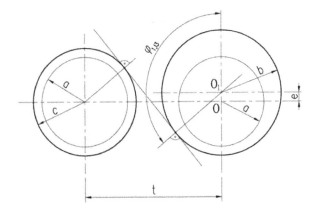

Figure 5. Limit angle $\varphi_{1,13}$

The limit angle $\varphi_{1,13}$ (Figure 5) can be expressed as

$$\varphi_{1,13} = \frac{\pi}{2} + \omega + \kappa,$$ (10)

where the angles κ i ω are given by

$$\kappa = \arctan \frac{b-c}{t},$$ (11)

$$\omega = \arccos \frac{b+c}{\sqrt{t^2 + e^2}}.$$ (12)

The view factor ψ in the interval $\varphi_{1,12} \leq \varphi_1 \leq \varphi_{1,13}$ is calculated from the following expression (Figure 6)

$$\psi = \frac{1}{2}\left(\sin \delta_2 - \sin \delta_1\right), \quad \varphi_{1,12} \leq \varphi_1 \leq \varphi_{1,13},$$ (13)

where

$$\delta_1 = \frac{\pi}{2},$$ (14)

$$\delta_2 = \varepsilon + \varphi_1 - \frac{\pi}{2},$$ (15)

$$\varepsilon = \beta + \gamma - \frac{\pi}{2}$$ (16)

$$\beta = \arcsin \frac{c}{\sqrt{\left(t-x_i\right)^2 + \left(y_i+e\right)^2}}, \tag{17}$$

$$\gamma = \pi - \arcsin \frac{t-x_i}{\sqrt{\left(t-x_i\right)^2 + \left(y_i+e\right)^2}}, \tag{18}$$

$$x_i = b\sin\varphi_1, \tag{19}$$

$$y_i = b\cos\varphi_1. \tag{20}$$

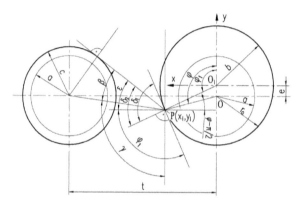

Figure 6. Determination of view factor in the angle interval $\varphi_{1,l2} \leq \varphi_1 \leq \varphi_{1,l3}$

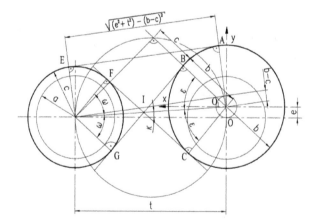

Figure 7. Determination of mean view factor ψ_{bs} for boiler setting over tube pitch t using the crossed string method

Radiation leaving the flame reaches also the boiler setting. The view factor for the radiation heat exchange between boiler setting and rear side of the measuring tube can be calculated in similar way as for the forward part. The mean heat flux q_{bs} resulting from the radiation heat transfer between the flame and the boiler setting can be determined using the crossed-string method [20-21].

The mean value of the view factor ψ_{bs} over the pitch length t is calculated from (Figure7)

$$\psi_{bs} = \frac{1}{2t}\left[(FC + BG) - (FG + BC)\right] \tag{21}$$

After substituting the lengths of straight FC and BG and circular segments FG and BC into Eq. (21), the mean value of the view factor over the boiler setting can be expressed as:

$$\psi_{bs} = \frac{b+c}{t}(\tan\omega - \omega). \tag{22}$$

The mean heat flux over the setting surface is

$$q_{bs} = q_m \psi_{bs}. \tag{23}$$

The angle ω is determined from

$$\tan\omega = \frac{\sqrt{e^2 + t^2 - (b+c)^2}}{b+c}, \tag{24}$$

If the diameters of the heat flux and water wall tubes are equal, then Eq.(24) simplifies to

$$\tan\omega = \sqrt{\left(\frac{t}{2c}\right)^2 - 1}. \tag{25}$$

The view factor for the radiation heat exchange between boiler setting and rear side of the measuring tube can be calculated in similar way as for the forward part. The view factor in the angle interval $\varphi_{1,l4} \le \varphi_1 \le \varphi_{1,l5}$ (Figure 8), accounting for the setting radiation, is given by

$$\psi = \psi_{bs} \cdot \frac{1}{2}(\sin\delta_2 - \sin\delta_1),\ \varphi_{1,l4} \le \varphi_1 \le \varphi_{1,l5} \tag{26}$$

where the limit angle $\varphi_{1,l4}$ is (Figure 8)

$$\psi_{1,l4} = \frac{\pi}{2} - \omega + \kappa. \tag{27}$$

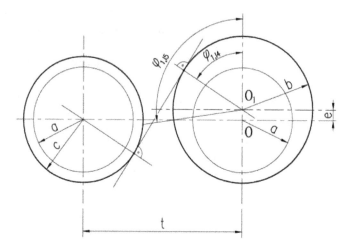

Figure 8. Limit angles $\varphi_{1,l4}$ and $\varphi_{1,l5} = \varphi_{1,l2} = \left(\pi/2\right) + \arcsin\left(e/b\right)$

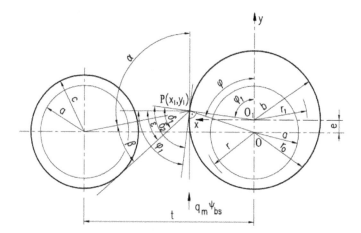

Figure 9. Determination of view factor in the angle interval $\varphi_{1,l4} \le \varphi_1 \le \varphi_{1,l5}$

The angles δ_1 and δ_2 are (Figure 9)

$$\delta_1 = \frac{\pi}{2} + \varepsilon - \varphi_{1'} \tag{28}$$

$$\delta_2 = \frac{\pi}{2}, \tag{29}$$

where

$$\varepsilon = \beta + \gamma - \frac{\pi}{2},$$

(30)

$$\beta = \arcsin \frac{c}{\sqrt{(t - x_i)^2 + (y_i + e)^2}},$$

(31)

$$\gamma = \pi - \arcsin \frac{t - x_i}{\sqrt{(t - x_i)^2 + (y_i + e)^2}},$$

(32)

$$x_i = b \sin \varphi_1,$$

(33)

$$y_i = b \cos \varphi_1.$$

(34)

The view factor ψ in the interval $\varphi_{1,l5} \le \varphi \le \pi$, where $\varphi_{1,l5} = \varphi_{1,l2}$, is given by

$$\psi = \psi_{bs} \cdot \frac{1}{2} \left(\sin \delta_1 + \sin \delta_2 \right), \ \varphi_{1,l5} \le \varphi \le \pi,$$

(35)

where

$$\delta_1 = \varphi_1 - \varepsilon - \frac{\pi}{2},$$

(36)

$$\delta_2 = \frac{\pi}{2},$$

(37)

$$\varepsilon = \frac{\pi}{2} - (\gamma - \beta),$$

(38)

$$\beta = \arcsin \frac{c}{\sqrt{(t - x_i)^2 + (y_i + e)^2}},$$

(39)

$$\gamma = \pi - \arcsin \frac{t - x_i}{\sqrt{(t - x_i)^2 + (y_i + e)^2}},$$

(40)

$$x_i = b \cos \left(\varphi_1 - \frac{\pi}{2} \right),$$

(41)

$$y_i = -b \sin \left(\varphi_1 - \frac{\pi}{2} \right).$$

(42)

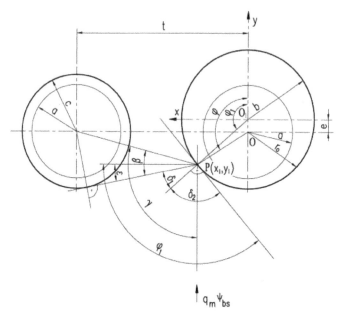

Figure 10. Determination of view factor in the angle interval $\varphi_{1,15} \leq \varphi_1 \leq \pi$

The total view factor accounts for the radiation heat exchange between the heat flux tube and flame and between the heat flux tube and the boiler setting.

2.2. Theory of the inverse problem

At first, the temperature distribution at the cross section of the measuring tube will be determined, i.e. the direct problem will be solved. Linear direct heat conduction problem can be solved using an analytical method. The temperature distribution will also be calculated numerically using the finite element method (FEM). In order to show accuracy of a numerical approach, the results obtained from numerical and analytical methods will be compared. The following assumptions have been made:

- thermal conductivity of the flux tube material is constant,
- heat transfer coefficient at the inner surface of the measuring tube does not vary on the tube circumference,
- rear side of the water wall, including the measuring tube, is thermally insulated,
- diameter of the eccentric flux tube is larger than the diameter of the water wall tubes,
- the outside surface of the measuring flux tube is irradiated by the flame, so the heat absorption on the tube fire side is non-uniform.

The cylindrical coordinate system is shown in Figure11.

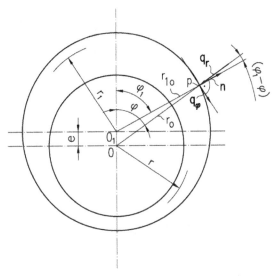

Figure 11. Approximation of the boundary condition on the outer tube surface

The temperature distribution in the eccentric heat flux tube is governed by heat conduction

$$\frac{1}{r}\frac{\partial}{\partial r}\left(kr\frac{\partial \theta}{\partial r}\right)+\frac{1}{r}\frac{\partial}{\partial \varphi}\left(\frac{k}{r}\frac{\partial \theta}{\partial \varphi}\right)=0 \tag{43}$$

subject to the following boundary conditions

$$k\nabla \theta \cdot \mathbf{n}\big|_{r=r_o} = q_m \psi\left(\varphi\right) \tag{44}$$

$$k\frac{\partial \theta}{\partial r}\bigg|_{r=a} = h\theta\big|_{r=a} \tag{45}$$

The left side of Eq. (44) can be transformed as follows (Figure 11)

$$k\nabla \theta \cdot \mathbf{n}\big|_{r=r_o} = \left(\mathbf{q_r}+\mathbf{q_\varphi}\right)\cdot \mathbf{n}\big|_{r=r_o} =$$
$$= \left[k\frac{\partial T}{\partial r}\cos\left(\varphi_1-\varphi\right)+\frac{k}{r}\frac{\partial T}{\partial \varphi}\sin\left(\varphi_1-\varphi\right)\right]\bigg|_{r=r_o} \tag{46}$$

The second term in Eq. (46) can be neglected since it is very small and the boundary condition (44) simplifies to

$$k\frac{\partial \theta}{\partial r}\bigg|_{r=r_o} = \frac{q_m\psi\left(\varphi\right)}{\cos\left(\varphi_1-\varphi\right)} \tag{47}$$

The heat flux over the tube circumference can be approximated by the Fourier polynomial

$$\frac{q_m \psi(\varphi)}{\cos(\varphi_1 - \varphi)} = q_0 + \sum_{n=1}^{\infty} q_n \cos(n\varphi) \tag{48}$$

where

$$q_0 = \frac{1}{\pi} \int_0^\pi \frac{q_m \psi(\varphi)}{\cos(\varphi_1 - \varphi)} d\varphi,$$

$$q_n = \frac{2}{\pi} \int_0^\pi \frac{q_m \psi(\varphi)}{\cos(\varphi_1 - \varphi)} \cos(n\varphi) d\varphi, \quad n = 1, \ldots \tag{49}$$

The boundary value problem (43, 45, 47) was solved using the separation of variables to give

$$\theta(r, \varphi) = A_0 + B_0 \ln r + \sum_{n=1}^{\infty} \left(C_n r^n + D_n r^{-n} \right) \cos n\varphi. \tag{50}$$

where

$$A_0 = \frac{q_0 r_o(\varphi)}{k} \left(\frac{1}{Bi} - \ln a \right), \tag{51}$$

$$B_0 = \frac{q_0 r_o(\varphi)}{k}, \tag{52}$$

$$C_n = \frac{q_n r_o(\varphi)}{k} \frac{\frac{1}{n} u^n (Bi + n) \frac{1}{a^n}}{Bi(u^{2n} + 1) + n(u^{2n} - 1)}, \tag{53}$$

$$D_n = -\frac{q_n r_o(\varphi)}{k} \frac{\frac{1}{n} u^n (Bi - n) a^n}{Bi(u^{2n} + 1) + n(u^{2n} - 1)}. \tag{54}$$

The ratio of the outer to inner radius of the eccentric flux tube: $u = u(\phi) = r_o(\phi)/a$ depends on the angle ϕ, since the outer radius of the tube flux

$$r_o = e \cos\varphi + \sqrt{b^2 - (e \sin\varphi)^2} \tag{55}$$

is the function of the angle ϕ.

Eq. (50) can be used for the temperature calculation when all the boundary conditions are known. In the inverse heat conduction problem three parameters are to be determined:

- absorbed heat flux referred to the projected furnace wall surface: $x_1 = q_m$,
- heat transfer coefficient on the inner surface of the boiler tube: $x_2 = h$,
- fluid bulk temperature: $x_3 = T_f$.

These parameters appear in boundary conditions (44) and (45) and will be determined based on the wall temperature measurements at m internal points (r_i, φ_i)

$$T(r_i, \varphi_i) = f_i, \ i = 1, ..., m, \ m \geq 3. \tag{56}$$

In a general case, the unknown parameters: $x_1, ..., x_n$ are determined by minimizing sum of squares

$$S = (\mathbf{f} - \mathbf{T_m})^T (\mathbf{f} - \mathbf{T_m}), \tag{57}$$

where $\mathbf{f} = (f_1, ..., f_m)^T$ is the vector of measured temperatures, and $\mathbf{T_m} = (T_1, ..., T_m)^T$ the vector of computed temperatures $T_i = T(r_i, \varphi_i)$, $i = 1, ..., m$.

The parameters $x_1 ... x_n$, for which the sum (34) is minimum are determined using the Levenberg-Marquardt method [23,25]. The parameters, $\mathbf{x}$, are calculated by the following iteration

$$\mathbf{x}^{(k+1)} = \mathbf{x}^{(k)} + \delta^{(k)}, \ k = 0, 1, ... \tag{58}$$

where

$$\delta^{(k)} = \left[\left(\mathbf{J}_m^{(k)} \right)^T \mathbf{J}_m^{(k)} + \mu^{(k)} \mathbf{I}_n \right]^{-1} \times$$
$$\times \left(\mathbf{J}_m^{(k)} \right)^T \left[\mathbf{f} - \mathbf{T}_m \left(\mathbf{x}^{(k)} \right) \right]. \tag{59}$$

where $\mu^{(k)}$ is the multiplier and $\mathbf{I}_n$ is the identity matrix. The Levenberg–Marquardt method is a combination of the Gauss–Newton method ($\mu^{(k)} \to 0$) and the steepest-descent method ($\mu^{(k)} \to \infty$). The $m \times n$ Jacobian matrix of $T(\mathbf{x}^{(k)}, r_i)$ is given by

$$\mathbf{J}^{(k)} = \frac{\partial \mathbf{T}(\mathbf{x})}{\partial \mathbf{x}_T} \bigg|_{\mathbf{x}=\mathbf{x}^{(k)}} = \begin{bmatrix} \dfrac{\partial T_1}{\partial x_1} & \cdots & \dfrac{\partial T_1}{\partial x_n} \\ \cdots & \cdots & \cdots \\ \cdots & \cdots & \cdots \\ \cdots & \cdots & \cdots \\ \cdots & \cdots & \cdots \\ \dfrac{\partial T_m}{\partial x_1} & \cdots & \dfrac{\partial T_m}{\partial x_n} \end{bmatrix}_{\mathbf{x}=\mathbf{x}^{(k)}}, \ m = 5, \ n = 3, \tag{60}$$

The symbol I_n denotes the identity matrix of $n \times n$ dimension, and $\mu^{(k)}$ the weight coefficient, which changes in accordance with the algorithm suggested by Levenberg and Marquardt. The upper index T denotes the transposed matrix. Temperature distribution $T(r, \varphi, \mathbf{x}^{(k)})$ is computed at each iteration step using Eq. (50). After a few iterations we obtain a convergent solution.

2.3. The uncertainty of the results

The uncertainties of the determined parameters $\mathbf{x}^*$ will be estimated using the error propagation rule of Gauss [23-26]. The propagation of uncertainty in the independent variables: measured wall temperatures f_j, j=1, ...m, thermal conductivity k, radial and angular positions of temperature sensors r_j, φ_j, j=1, ...m is estimated from the following equation

$$2\sigma_{x_i} = \left[\sum_{j=1}^{m} \left(\frac{\partial x_i}{\partial f_j} \sigma_{f_j} \right)^2 + \left(\frac{\partial x_i}{\partial k} \sigma_k \right)^2 + \sum_{j=1}^{m} \left(\frac{\partial x_i}{\partial r_j} \sigma_{r_j} \right)^2 + \sum_{j=1}^{m} \left(\frac{\partial x_i}{\partial \varphi_j} \sigma_{\varphi_j} \right)^2 \right]^{1/2}, \tag{61}$$

$$i = 1, 2, 3$$

The 95% uncertainty in the estimated parameters can be expressed in the form

$$x_i = x_i^* \pm 2\sigma_{x_i}, \tag{62}$$

where $x_i^*, i = 1, 2, 3$ represent the value of the parameters obtained using the least squares method. The sensitivity coefficients $\partial x_i / \partial f_j$, $\partial x_i / \partial k$, $\partial x_i / \partial r_j$, and $\partial x_i / \partial \varphi_j$ in the expression (61) were calculated by means of the numerical approximation using central difference quotients:

$$\frac{\partial x_i}{\partial f_j} = \frac{x_i \left(f_1, f_2, ..., f_j + \delta, ..., f_m \right) - x_i \left(f_1, f_2, ..., f_j - \delta, ..., f_m \right)}{2\delta}, \tag{63}$$

where δ is a small number.

2.4. Computational and boiler tests

Firstly, a computational example will be presented. "Experimental data" are generated artificially using the analytical solution (50).

Consider a water-wall tube with the following parameters (Figure1.):

- outer radius b = 35 mm,
- inner radius a = 25 mm,
- pitch of the water-wall tubes t = 80 mm,
- thermal conductivity k = 28.5 W/(m·K),

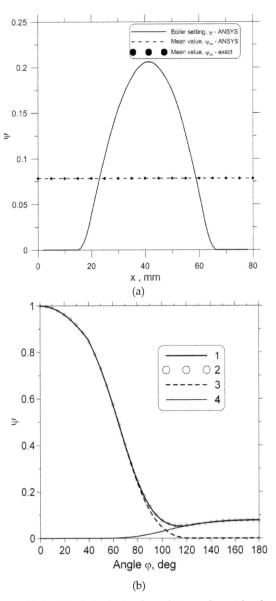

Figure 12. View factor associated with radiation heat exchange between elemental surface on the boiler setting or flux tube and flame: (a) – view factor for radiation heat transfer between flame and boiler setting, (b) 1 - total view factor accounting radiation from furnace and boiler setting, 2 - approximation by the Fourier polynomial of the seventh degree, 3 - exact view factor for furnace radiation, 4- view factor from boiler setting

- absorbed heat flux $q_m = 200000$ W/m²,
- heat transfer coefficient $h = 30000$ W/(m²·K),
- fluid temperature $T_f = 318$ °C.

The view factor distributions on the outer surface of the flux-tube and boiler setting were calculated analytically and numerically by means of the finite element method (FEM) [22]. The changes of the view factor over the pitch length and tube circumference are illustrated in Figures 12 and 13.

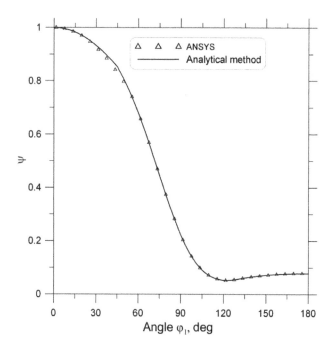

Figure 13. Comparison of total view factor calculated by exact and FEM method

The agreement between the temperatures of the outer and inner tube surfaces which were calculated analytically and numerically is also very good (Figures 14 and 15). The small differences between the analytical and FEM solutions are caused by the approximate boundary condition (47). The temperature distribution in the flux tube cross section is shown in Figure 14.

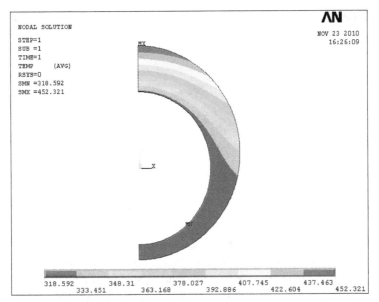

Figure 14. Computed temperature distribution in °C in the cross section of the heat flux tube; q_m = 200000 W/m², h = 30000 W/(m²·K), T_f=318 °C

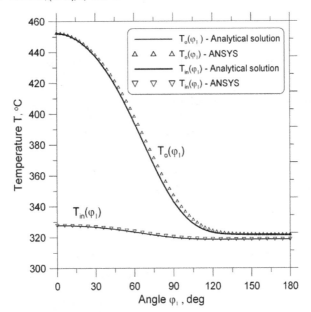

Figure 15. Temperature distribution at the inner and outer surfaces of the flux tube calculated by the analytical and finite element method

The following input data is generated using Eq. (50): $f_1 = 437.98 \, ^\circ C$, $f_2 = 434.47 \, ^\circ C$, $f_3 = 383.35 \, ^\circ C$, $f_4 = 380.70 \, ^\circ C$, $f_5 = 321.58 ^\circ C$.

The following values were obtained using the proposed method: $q_m^* = 200\,000.35 \, \text{W/m}^2$, $h^* = 30\,001.56 \, \text{W/(m}^2 \cdot \text{K)}$, $T_f^* = 318.00 \, ^\circ C$.

In order to show the influence of the measurement errors on the determined thermal boundary parameters, the 95% confidence intervals were calculated. The following uncertainties of the measured values were assumed (at a 95% confidence interval):

$$2\sigma_{f_j} = \pm 0.2 \text{K}, j = 1, \ldots, 5, 2\sigma_k = \pm 0.5 \, \text{W}/(\text{m} \cdot \text{K}), 2\sigma_{r_j} = \pm 0.05 \text{mm}, 2\sigma_{\varphi_j} = \pm 0.5^\circ, j = 1, \ldots, 5.$$

The uncertainties (95% confidence interval) of the coefficients x_i were determined using the error propagation rule formulated by Gauss.

The calculation using Eq. (61) yielded the following results: $x_1 = 200\,000.35 \pm 3827.72 \, \text{W/m}^2$, $x_2 = 30\,001.56 \pm 2698.81 \, \text{W/(m}^2 \cdot \text{K)}$, $x_3 = 318.0 \pm 0.11 \, ^\circ C$. The accuracy of the obtained results is very satisfactory. There is only a small difference between the estimated parameters and the input values. The highest temperature occurs at the crown of the flux-tube (Figures 14 and 15). The temperature of the inner surface of the flux tube is only a few degrees above the saturation temperature of the water-steam mixture. Since the heat flux at the rear side of the tube is small, the circumferential heat flow rate is significant. However, the rear surface thermocouple indicates temperatures of 2-4 °C above the saturation temperature. Therefore, the fifth thermocouple can be attached to the unheated side of the tube so as to measure the temperature of the water-steam mixture flowing through the flux tube.

In the second example, experimental results will be presented. Measurements were conducted at a 50MW pulverized coal fired boiler. The temperatures indicated by the flux tube at the elevation of 19.2 m are shown in Figure 16. The heat flux tube is of 20G low carbon steel with temperature dependent thermal conductivity

$$k(T) = 53.26 - 0.02376224T, \tag{64}$$

where the temperature T is expressed in °C and thermal conductivity in W/(m·K).

The unknown parameters were determined for eight time points which are marked in Figure 16.

The inverse analysis was performed assuming the constant thermal conductivity $k(\bar{T})$ which was obtained from Eq.(64) for the average temperature: $\bar{T} = (T_1 + T_2 + T_3 + T_4)/4$.

The estimated parameters: heat flux q_m, heat transfer coefficient h, and the water-steam mixture T_f are depicted in Figure 17. The developed flux tube can work for a long time in the destructive high temperature atmosphere of a coal-fired boiler.

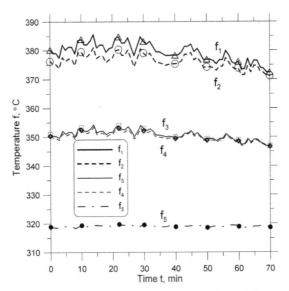

Figure 16. Measured flux tube temperatures; marks denote measured temperatures taken for the inverse analysis

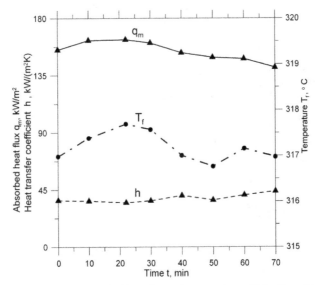

Figure 17. Estimated parameters: absorbed heat flux q_m, heat transfer coefficient h, and temperature of water-steam mixture T_f

Flux tubes can also be used as a local slag monitor to detect a build up of slag. The presence of the scale on the inner surface of the tube wall can also be detected.

3. Tubular type heat flux meter made of a finned tube

In this section, a numerical method for determining the heat flux in boiler furnaces, based on experimentally acquired interior flux-tube temperatures, is presented. The tubular type instrument has been designed (Figure 18) to provide a very accurate measurement of absorbed heat flux q_m, inside heat transfer coefficient h_{in}, and water steam temperature T_f. The number of thermocouples is greater than three because the additional information can help enhance the accuracy of parameter determining. In contrast to the existing devices, in the developed flux-tube fins are not welded to adjacent water-wall tubes. Temperature distribution in the flux-tube is symmetric and not disturbed by different temperature fields in neighboring tubes. The temperature dependent thermal conductivity of the flux-tube material was assumed. The meter is constructed from a short length of eccentric tube containing four thermocouples on the fire side below the inner and outer surfaces of the tube. The fifth thermocouple is located at the rear of the tube (on the casing side of the water-wall tube). The boundary conditions on the outer and inner surfaces of the water flux-tube must then be determined from temperature measurements in the interior locations. Four K-type sheathed thermocouples, 1 mm in diameter, are inserted into holes, which are parallel to the tube axis. The thermal conduction effect at the hot junction is minimized because the thermocouples pass through isothermal holes. The thermocouples are brought to the rear of the tube in the slot machined in the tube wall. An austenitic cover plate with the thickness of 3 mm – welded to the tube – is used to protect the thermocouples from the incident flame radiation. A K-type sheathed thermocouple with a pad is used to measure the temperature at the rear of the flux-tube. This temperature is almost the same as the water-steam temperature. An inverse problem of heat conduction was solved using the least squares method. Three unknown parameters were estimated using the Levenberg-Marquardt method [23, 25]. At every iteration step, the temperature distribution over the cross-section of the heat flux meter was computed using the ANSYS CFX software

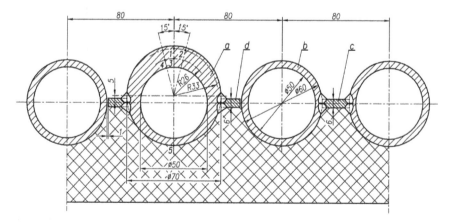

Figure 18. The cross-section of the membrane wall in the combustion chamber of the steam boiler

Test calculations were carried out to assess accuracy of the presented method. The uncertainty in determined parameters was calculated using the Gauss variance propagation rule. The presented method is appropriate for membrane water walls (Figure 18). The new method has advantages in terms of simplicity and flexibility.

3.1. Theory

The furnace wall tubes in most modern units are welded together with steel bars (fins) to provide membrane wall panels which are insulated on one side and exposed to a furnace on the other, as shown schematically in Figure 18.

In a heat conduction model of the flux-tube the following assumptions are made:

- temperature distribution is two-dimensional and steady-state,
- the thermal conductivity of the flux-tube and membrane wall,
- may be dependent of temperature,
- the heat transfer coefficient h_{in} and the scale thickness ds is uniform over the inner tube surface.

The temperature distribution is governed by the non-linear partial differential equation

$$\nabla \cdot \left[k(T)\nabla T \right] = 0, \tag{65}$$

where ∇ is the vector operator, which is called nabla (gradient operator), and in Cartesian coordinates is defined by $\nabla = i\partial/\partial x + j\partial/\partial y + k\partial/\partial z +$. The unknown boundary conditions may be expressed as

$$\left[k(T)\frac{\partial T}{\partial n} \right]_{s} = q(s), \tag{66}$$

where $q(s)$ is the radiation heat flux absorbed by the exposed flux tube and membrane wall surface. The local heat flux $q(s)$ is a function of the view factor $\psi(s)$ (Figure 19)

$$q(s) = q_m \psi(s), \tag{67}$$

where q_m is measured heat flux (thermal loading of heating surface). The view factor $\psi(s)$ from the infinite flame plane to the differential element on the membrane wall surface can be determined graphically [7], or numerically [22].

In this chapter, $\psi(s)$ was evaluated numerically using the finite element program ANSYS [22], and is displayed in Figure 19 as a function of the extended coordinate s. Because of the symmetry, only the representative water-wall section illustrated in Figure 20 needs to be analyzed. The convective heat transfer from the inside tube surfaces to the water-steam mixture is described by Newton's law of cooling

$$-\left[k(T)\frac{\partial T}{\partial n}\right]_{s_{in}} = h_{in}\left(T\big|_{s_{in}} - T_f\right),\tag{68}$$

where $\partial T/\partial n$ is the derivative in the normal direction, h_{in} is the heat transfer coefficient and T_f denotes the temperature of the water–steam mixture.

The reverse side of the membrane water-wall is thermally insulated. In addition to the unknown boundary conditions, the internal temperature measurements f_i are included in the analysis

$$T_e(\mathbf{r}_i) = f_i, \quad i = 1,\dots,m,\tag{69}$$

where $m = 5$ denotes the number of thermocouples (Figure 18). The unknown parameters: $x_1 = q_m$, $x_2 = h_{in}$, and $x_3 = T_f$ were determined using the least-squares method. The symbol r_{in} denotes the inside tube radius, and $k(T)$ is the temperature dependent thermal conductivity. The object is to choose $\mathbf{x} = (x_1, \dots, x_n)^T$ for $n = 3$ such that computed temperatures $T(\mathbf{x}, \mathbf{r}_i)$ agree within certain limits with the experimentally measured temperatures f_i.

This may be expressed as

$$T(\mathbf{x}, \mathbf{r}_i) - f_i \cong 0, \quad i = 1,\dots,m, \quad m = 5.\tag{70}$$

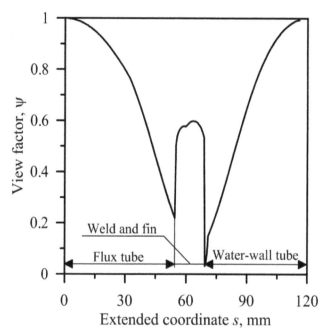

Figure 19. View factor distribution on the outer surface of water-wall tube

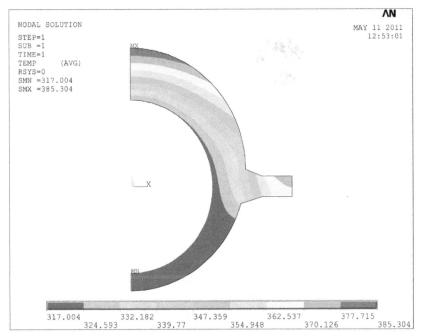

Figure 20. Temperature distribution in the flux tube cross-section for: q_m = 150000 W/m², h_{in} = 27000 W/(m²·K) and T_f = 317°C

The least-squares method is used to determine parameters **x**. The sum of squares

$$S = \sum_{i=1}^{m} \left[f_i - T(\mathbf{x}, \mathbf{r}_i) \right]^2, \quad m = 5, \tag{71}$$

is minimized using the Levenberg–Marquardt method [23, 25].

The uncertainties of the determined parameters **x*** will be estimated using the error propagation rule of Gauss [23-26].

3.2. Test computations

The flux-tubes were manufactured in the laboratory and then securely welded to the water-wall tubes at different elevations in the furnace of the steam boiler. The coal fired boiler produces 58.3 kg/s superheated steam at 11 MPa and 540°C.

The material of the heat flux-tube is 20G steel. The composition of the 20G mild steel is as follows: 0.17–0.24% C, 0.7–1.0% Mn, 0.15–0.40% Si, Max 0.04% P, Max 0.04% S, and the remainder is iron Fe. The heat flux-tube thermal conductivity is assumed to be temperature dependent (Table 1).

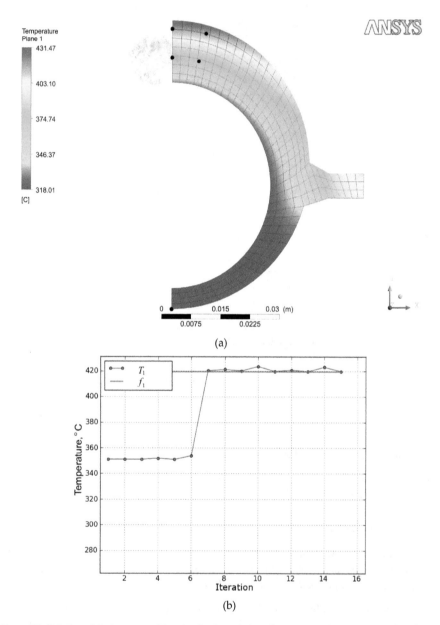

(a)

(b)

Figure 21. Solution of the inverse problem for the "exact" data: $f_1 = 419.66°C$, $f_2 = 417.31°C$, $f_3 = 374.90°C$, $f_4 = 373.19°C$, $f_5 = 318.01°C$; (a) - temperature distribution in the flux-tube, (b) - iteration number for the temperature T_1

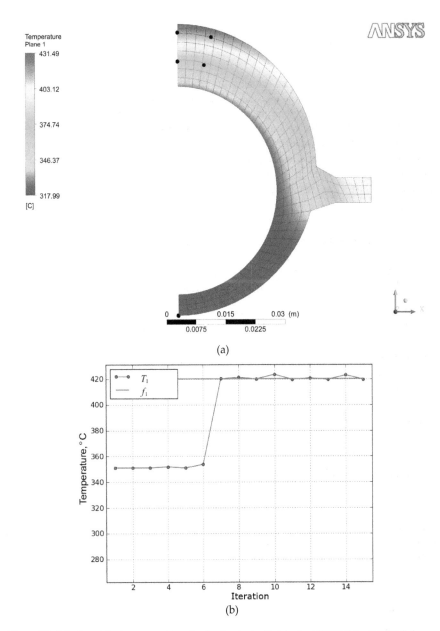

(a)

(b)

Figure 22. Solution of the inverse problem for the "perturbed" data: f_1 = 420.16°C, f_2 = 416.81°C, f_3 = 375.40°C, f_4 = 372.69°C, f_5 = 318.01°C; (a) - temperature distribution in the flux-tube , (b) - iteration number for the temperature T_1

Temperature T, °C	100	200	300	400
Thermal conductivity k, W/(m·K)	50.69	48.60	46.09	42.30

Table 1. Thermal conductivity $k(T)$ of steel 20G as a function of temperature

To demonstrate that the maximum temperature of the fin tip is lower than the allowable temperature for the 20G steel, the flux tube temperature was computed using ANSYS CFX package [22]. Changes of the view factor on the flux tube, weld and fin surface were calculated with ANSYS CFX. The temperature distribution shown in Figure 20 was obtained for the following data: absorbed heat flux, q_m = 150000 W/m², temperature of the water-steam mixture, T_f = 317°C, and heat transfer coefficient at the tube inner surface, h_{in} = 27000 W/(m²·K). An inspection of the results shown in Figure 20 indicates that the maximum temperature of the fin does not exceed 375°C.

Next, to illustrate the effectiveness of the presented method, test calculations were carried out. The "measured" temperatures f_i, i = 1, 2, ..., 5 were generated artificially by means of ANSYS CFX for: q_m = 250000 W/m², h_{in} = 30000 W/(m²·K) and T_f = 318°C. The following values of "measured" temperatures were obtained f_1 = 419.66°C, f_2 = 417.31°C, f_3 = 374.90°C, f_4 = 373.19°C, f_5 = 318.01°C. The temperature distribution in the flux tube cross-section, reconstructed on the basis of five measured temperatures is depicted in Figure 21a.

The proposed inverse method is very accurate since the estimated parameters: q_m = 250000.063 W/m², h_{in} = 30000.054 W/(m² ·K) and T_f = 318.0°C differ insignificantly from the input values. In order to show the influence of the measurement errors on the determined parameters, the 95% confidence intervals were estimated. The following uncertainties of the measured values were assumed (at 95% confidence interval): $2\sigma_{f_j}$ = ±0.5 K, j = 1, 2, ..., 5,

$2\sigma_k = \pm 1$ W/$(m \cdot K)$, $2\sigma_{r_j}$ = ±0.05mm, $2\sigma_{\varphi_j}$ = ±0.5°, j=1,...,5. The uncertainties (95% confidence interval) of the coefficients x_i were determined using the error propagation rule formulated by Gauss [23-26]. The calculated uncertainties are: ±6% for q_m, ±33% for h_{in} and ±0.3% for T_f. The accuracy of the results obtained is acceptable.

Then, the inverse analysis was carried out for perturbed data: f_1 = 420.16°C, f_2 = 416.81°C, f_3 = 375.40°C, f_4 = 372.69°C, f_5 = 318.01°C. The reconstructed temperature distribution illustrates Figure 22a.

The obtained results are: q_m = 250118.613 W/m², h_{in} = 30050.041 W/(m² ·K) and T_f = 317.99°C. The errors in the measured temperatures have little effect on the estimated parameters. The number of iterations in the Levenberg-Marquardt procedure is small in both cases (Figures 21b and 22b).

4. Conclusions

Two different tubular type instruments (flux tubes) were developed to identify boundary conditions in water wall tubes of steam boilers. The first measuring device is an eccentric tube. The ends of the four thermocouples are located at the fireside part of the tube and the fifth thermocouple is attached to the unheated rear surface of the tube. The meter presented in the paper has one particular advantage over the existing flux tubes to date. The temperature distribution in the flux tube is not affected by the water wall tubes, since the flux tube is not connected to adjacent waterwall tubes with metal bars, referred to as membrane   or webs. To determine the unknown parameters only the temperature distribution at the cross section of the flux tube must be analyzed.

The second flux tube has two longitudinal fins. Fins attached to the flux tube are not welded to the adjacent water-wall tubes, so the temperature distribution in the measuring device is not affected by neighboring water-wall tubes. The installation of the flux tube is easier because welding of fins to adjacent water-wall tubes is avoided. Based on the measured flux tube temperatures the non-linear inverse heat conduction problem was solved. A CFD based method for determining heat flux absorbed water wall tubes, heat transfer coefficient at the inner flux tube surface and temperature of the water-steam mixture has been presented. The proposed flux tube and the inverse procedure for determining absorbed heat flux can be used both when the inner surface of the heat flux tube is clean and when scale or corrosion deposits are present on the inner surface what can occur after a long time service of the heat flux tube.

The flux tubes can work for a long time in the destructive high temperature atmosphere of a coal-fired boiler.

Nomenclature

a	inner radius of boiler tube and flux-tube (m)
b	outer radius of flux-tube (m)
Bi	Biot number, $Bi = ha/k$
c	outer radius of boiler tube (m)
e	eccentric (m)
f_i	measured wall temperature at the i-th location (^{o}C or K)
$\mathbf{f}$	vector of measured wall temperatures
h	heat transfer coefficient (W/(m^2·K))
$\mathbf{I}_n$	identity matrix
$\mathbf{J}$	Jacobian matrix of $\mathbf{T}$
k	thermal conductivity (W/(m·K))
l	arbitrary length of boiler tube (m)
m	number of temperature measurement points
n	number of unknown parameters
q_m	heat flux to be determined (absorbed heat flux referred to the projected furnace water wall surface) (W/m^2)
r	coordinate in cylindrical coordinate system or radius (m)
r_i	radial coordinate of the i-th thermocouple (m)
r_{in}	inner radius of the flux-tube (m)
r_o	outer radius of the flux-tube (m)

r	position vector
s	extended coordinate along the fireside water-wall surface (m)
S	sum of the temperature difference squares (K^2)
t	pitch of the water wall tubes (m)
T	temperature (°C or K)
T_f	fluid temperature (°C or K)
T_i	calculated temperature at the location(r_i,ϕ_i) (°C or K)
T^m	m - dimensional column vector of calculated temperatures
$u\,(\phi)$	ratio of the outer to the inner radius of the tube, $u\,(\phi) = r_o/a$
x_i	unknown parameter
x	n-dimensional column vector of unknown parameters

Greek symbols

$\alpha, \beta, \gamma, \delta_1, \delta_2, \varepsilon$	angles (rad)
θ	temperature excess over the fluid temperature, $\theta = T - T_f$ (K)
ϕ	angular coordinate (rad)
ϕ_i	angular coordinate of the i-th thermocouple (rad)
μ	multiplier in the Levenberg-Marquardt algorithm
ψ	view factor

Subscripts

in	inner
o	outer
i	i-th temperature measurement point
f	fluid

Author details

Jan Taler
Department of Thermal Power Engineering, Cracow University of Technology, Cracow, Poland

Dawid Taler
Institute of Heat Transfer Engineering and Air Protection, Cracow University of Technology, Cracow, Poland

5. References

[1] Segeer M, Taler J (1983) Konstruktion und Einsatz transportabler Wärmeflußsonden zur Bestimmung der Heizflächenbelastung in Feuerräumen. Fortschr.-Ber. VDI Zeitschrift, Reihe 6, Nr 129. Düsseldorf : VDI-Verlag.

[2] Northover EW, Hitchcock JA (1967) A Heat Flux Meter for Use in Boiler Furnaces. J. Sci. Instrum. 44: 371–374.

[3]   Neal SBH, Northover EW (1980) The Measurement of Radiant Heat Flux in Large Boiler Furnaces-I. Problems of Ash Deposition Relating to Heat Flux. Int. J. Heat Mass Transfer 23: 1015–1022.

[4]   Arai N, Matsunami A, Churchill S (1996) A Review of Measurements of Heat Flux Density Applicable to the Field of Combustion. Exp. Therm. Fluid Sci. 12: 452–460.

[5]   Taler J (1990) Measurement of Heat Flux to Steam Boiler Membrane Water Walls. VGB Kraftwerkstechnik 70: 540–546.

[6]   Taler J (1992) A Method of Determining Local Heat Flux in Boiler Furnaces. Int. J. Heat Mass Transfer 35:1625–1634.

[7]   Taler J (1990) Messung der lokalen Heizflächenbelastung in Feuerräumen von Dampferzeugern. Brennstoff-Wärme-Kraft (BWK) 42: 269-277.

[8]   Fang Z, Xie D, Diao N, Grace JR, Lim CJ (1997) A New Method for Solving the Inverse Conduction Problem in Steady Heat Flux Measurement. Int. J. Heat Mass Transfer 40: 3947–3953.

[9]   Luan W, Bowen BD, Lim CJ, Brereton CMH, Grace JR (2000) Suspension-to Membrane-Wall Heat Transfer in a Circulating Fluidized Bed Combustor. Int. J. Heat Mass Transfer 43: 1173–1185.

[10] Taler J, Taler D (2007) Tubular Type Heat Flux Meter for Monitoring Internal Scale Deposits in Large Steam Boilers. Heat Transfer Engineering 28: 230-239.

[11] Sobota T, Taler D (2010) A Simple Method for Measuring Heat Flux in Boiler Furnaces. Rynek Energii 86: 108-114.

[12] Taler D, Taler J, Sury A (2011) Identification of Thermal Operation Conditions of Water Wall Tubes Using Eccentric Tubular Type Meters. Rynek Energii 92: 164-171.

[13] Taler J, Taler D, Kowal A (2011) Measurements of Absorbed Heat Flux and Water-side Heat Transfer Coefficient in Water Wall Tubes. Archives of Thermodynamics 32: 77 – 88.

[14] Taler J, Taler D, Sobota T, Dzierwa P (2011) New Technique of the Local Heat Flux Measurement in Combustion Chambers of Steam Boilers. Archives of Thermodynamics 32: 103-116.

[15] LeVert FE, Robinson JC, Frank RL, Moss RD, Nobles WC, Anderson AA (1987) A Slag Deposition Monitor for Use in Coal_Fired Boilers. ISA Transactions 26: 51-64

[16] LeVert FE, Robinson JC, Barrett SA, Frank RL, Moss RD, Nobles WC, Anderson AA (1988) Slag Deposition Monitor for Boiler Performance Enhancement. ISA Transactions 27: 51-57

[17] Vallero A, Cortes C (1996) Ash Fouling in Coal-Fired Utility Boilers. Monitoring and Optimization of On-Load Cleaning. Prog. Energy. Combust. Sci. 22: 189–200.

[18] Teruel E, Cortes C, Diez LI, Arauzo I (2005) Monitoring and Prediction of Fouling in Coal-Fired Utility Boilers Using Neural Networks. Chem. Eng. Sci. 60: 5035–5048.

[19] Taler J, Trojan M, Taler D (2011) Monitoring of Ash Fouling and Internal Scale Deposits in Pulverized Coal Fired Boilers. New York: Nova Science Publishers.

[20] Howell JR, Siegel R, Mengüç MP (2011) Thermal Radiation Heat Transfer. Boca Raton: CRC Press - Taylor & Francis Group.

[21] Sparrow FM, Cess RD (1978) Radiation Heat Transfer. New York: McGraw-Hill.

[22] ANSYS CFX 12. (2010) Urbana, Illinois, USA: ANSYS Inc.

[23] Seber GAF, Wild CJ (1989) Nonlinear regression. New York: Wiley.

[24] Policy on reporting uncertainties in experimental measurements and results (2000). ASME J. Heat Transfer 122: 411–413.

[25] Press WH, Teukolsky SA, Vetterling WT, Flannery BP (2006) Numerical Recipes in Fortran. The Art of Scientific Computing. Cambridge: Cambridge University Press.

[26] Coleman HW, Steele WG (2009) Experimentation, Validation, and Uncertainty Analysis for Engineers. Hoboken: Wiley.

Chapter 2

Analytical and Experimental Investigation About Heat Transfer of Hot-Wire Anemometry

Mojtaba Dehghan Manshadi and Mohammad Kazemi Esfeh

Additional information is available at the end of the chapter

1. Introduction

The hot-wire anemometer is a famous thermal instrument for turbulence measurements [1]. The principle of operation of the anemometer is based on the heat transfer from a fine filament where it is exposed to an unknown flow that varies with deviation in the flow rate. The hot-wire filament is made from a special material that processes a temperature coefficient of resistance [2]. Thermal anemometry is the most common method employed to measure instantaneous fluid velocity. It may be operated in one of these two modes, constant current (CC) mode and constant temperature (CT) mode.

Constant-Current (CC) mode: In this mode, the current flow through the hot wire is kept constant and variation in the wire resistance caused by the fluid flow is measured by monitoring the voltage drop variations across the filament.

Constant Temperature (CT) mode: In this mode, the hotwire filament is positioned in a feedback circuit and tends to maintain the hotwire at a constant resistance and hence at a constant temperature and fluctuations in the cooling of the hot wire, filaments are similar to variations in the current flow through the hotwire.

Hot wire anemometers are normally operated in the constant (CTA) mode. The hot-wire anemometry has been used for many years in fluid mechanics as a relatively economical and effective method of measuring the flow velocity and turbulence. It is based on the convective heat transfer from a heated sensing element .Briefly; any fluid velocity change would cause a corresponding change of the convective heat loss to the surrounding fluid from an electrically heated sensing probe. The variation of heat loss from the thermal element can be interpreted as a measure of the fluid velocity changes. In subsonic incompressible flow the heat transfer from a hot wire sensor is dependent on the mass flow, ambient temperature and wire temperature. Since density variations are assumed to be zero, the mass flow variations are only function of velocity changes. The major advantage of maintaining the hot wire at a constant operational temperature and thereby at a constant resistance is that the thermal inertia of the sensing element is automatically adjusted when the flow conditions are varied. The electronic circuit of chosen CTA is shown schematically in Fig. 1. This mode of operation is achieved by incorporating a feedback differential amplifier into the hot-wire anemometer circuit. Such set-up obtains a rapid variation in the heating current and compensates for instantaneous changes in the flow velocity [2]. The sensing element in case studied in this research is a tungsten wire that is heated by an electric current to a temperature of approximately 250 °C. The heat is transferred from the wire mainly through convection. This heat loss is strongly dependent on the excess temperature of the wire, the physical properties of the sensing element and its geometrical configuration. The authors strive to present an analytical solution for heat transfer equation of hotwire for states that can ignore the radiation term for the wire. The fundamental principle of hot-wire anemometer is based on the convective heat transfer, thus in the research, an attempt is made to develop a better perception from the heat transfer of the hot-wire sensor. Also, the effect of air flow temperature variations on the voltage of hot wire, CTA has been studied experimentally. Furthermore, on the basis of air flow velocity and ambient temperature variations, the percentage errors in velocity measurements have been calculated. Finally, based on results, an accurate method has been proposed to compensate for air flow temperature variations.

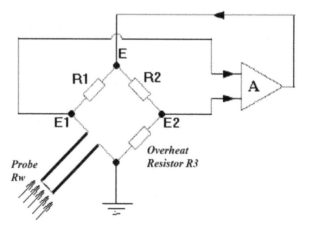

Figure 1. Schematic of a constant temperature anemometer

2. Theoretical background

The hot-wire involves one part of a Wheatstone bridge, where the wire resistance is kept constant over the bandwidth of the feedback loop. The electrical power dissipation $\dot{Q}_{elec}$, when the sensor is heated, is given by:

$$\dot{Q}_{elec} = I^2 R_w \tag{1}$$

I and R_w are the current passing through the sensor and the resistance of the sensor at the temperature Tw, respectively. The convection heat transfer rate to the fluid can be expressed in terms of the heat-transfer coefficient h, as:

$$\dot{Q} = A_w h(T_w - T_a) \tag{2}$$

Where A_w is the surface area of sensor and (T_w-T_a) is the difference between the temperature of the hot-wire sensor and the temperature of the fluid. For steady-state operation, the rate of electric power dissipation equals to the rate of convective heat transfer (assuming the conductive heat transfer to the two prongs is negligible). Thus,

$$I^2 R_w = \pi dLh(T_w - T_a) = \pi Lk(T_w - T_a)Nu \tag{3}$$

By introducing the wire voltage $E_w = IR_w$ and using equation (3), one can conclude that (k is the thermal conductivity of the fluid):

$$\frac{E_w^2}{R_w} = \pi Lk(T_w - T_a)Nu \tag{4}$$

According to the pioneering experimental and theoretical work by King, the convective heat transfer is often expressed in the following form:

$$Nu = A + BRe^n \tag{5}$$

Where A and B are empirical calibration constants. For long wires in air, King found that A=0.338, B=0.69 and n=0.5. It is interesting to note that King based his derivations on the assumption of potential flow, which is a poor approximation of the real flow around a wire at low Reynolds' numbers, so King's derivation is in a sense approximately erroneous. Nevertheless, King's law has been the considered tool for fitting calibration data in practical hot-wire anemometry for almost a hundred years [3].

By introducing equation (5) into equation (4) can give:

$$\frac{E_w^2}{R_w} = \pi Lk(T_w - T_a)(A + B\left(\frac{\rho d}{\mu}\right)^{0.5} U^{0.5}) \tag{6}$$

Equation (6) states that the hot-wire voltage is sensitive both to the velocity and temperature of air. Here, rearranging the equation (6) gives:

$$\frac{E_w^2}{R_w} = (A + BU^{0.5})(T_w - T_a) \tag{7}$$

Where π, l, k, d, ϱ and μ have been included in the constant coefficients A and B.

According to equation (7), Kanevce and Oka [4] introduced the following expression to correct the hot-wire output voltage for the temperature drift:

$$E_{corr} = E_w \left(\frac{T_w - T_a}{T_w - T_{a,r}} \right)^{0.5} \tag{8}$$

$T_{a,r}$ is ambient reference temperature during sensor calibration and T_a is ambient temperature during data acquisition where E_{corr} is corrected voltage. For a hot wire probe with a finite length active wire element, the conductive heat transfer to prongs must be taken into account. In practice this is often achieved by the modifying equation [7] as:

$$\frac{E_w^2}{R_w} = \left(A + BU^n \right) \left(T_w - T_a \right) \tag{9}$$

The values of A, B and n can be determined by a suitable calibration procedure. It should be noted that the term (T_w-T_a) and physical properties of fluid are dependent on the ambient temperature. In the previous related studies, the effect of term (T_w-T_a) is considered only to compensate the ambient temperature variations [5]. In other word, the variations of physical properties of fluid and Nusselt number are ignored. So in this study, the variations of Nusselt number with the fluid temperature have been considered. The following equation for correction of output voltage E. has been proposed by the relations extended in Ref. [6].

$$E_{corr} = E_w \left(\frac{T_w - T_{a,r}}{T_w - T_a} \right)^{0.5(1 \pm m)} \tag{10}$$

In Ref. [7], equation (10) is employed to correct the voltage of CTA output. Results showed that the required error correction factor (m) depends on whether the fluid temperature decreases or increases with respect to the calibration temperature of the CTA.

The CT mode velocity and temperature sensitivities corresponding to equation (9) are:

$$S_u = \frac{\partial E_w}{\partial U} = \frac{nBU^{n-1}}{2} \left[\frac{R_w \left(T_w - T_a \right)}{A + BU^n} \right]^{0.5} \tag{11}$$

$$S_\theta = \frac{\partial E_w}{\partial \theta} = \frac{-1}{2} \left[\frac{R_w \left(A - BU^n \right)}{T_w - T_a} \right]^{0.5} \tag{12}$$

Where θ is a small fluctuation in the fluid temperature. Equations (11) and (12) show that the value of S_u increases and the value of S_θ decreases by increasing the value of (T_w-T_a). A high over-heat ratio (R_w/R_a) is recommended for the measurement of velocity fluctuations [2]. In Ref. [7], it is stated that for an over-heat ratio of 1.4, the error incurred amounts to about 2.5% per degree Celsius temperature change. With the increase in the overheat ratio to 1.6, the error in CTA output is reduced to about 2% per degree Celsius temperature change.

The heat transfer process from a hot-wire sensor is usually expressed in a non-dimensional form where involve a relationship between the Nusselt number, the Reynolds number and the Prandtl number. The Nusselt number is usually assumed to be a function of Reynolds and Prandtl numbers and under most flow conditions, the Prandtl number is constant.

In hot-wire anemometry, the sizes of the sensing element are small, so that the Reynolds number of the flow is very low and the flow pattern over the sensor can be assumed to be symmetrical and quasi-steady. Due to the statement of the flow continuity, the mean free path of the particles is very much less than the diameter of the wire and conventional heat transfer theories are applicable [8]. Furthermore, the length of the sensor is much greater than its diameter. Hence, it may be assumed that the loss conduction through the ends is negligible and the relation for the heat transfer from an infinite cylinder can be applied. Kramers [9] has proposed the following equation based on heat-transfer experimental results for wires (with infinite length-to-diameter ratio), placed in air, water and oil:

$$Nu = 0.42Pr^{0.2} + 0.57Pr^{0.33}Re^{0.5} \qquad (13)$$

He selected the film temperature $T_f = (T_w+T_a)/2$ as the reference temperature for the fluid properties.

3. Experimental procedure

An air condition unit was used to carry out the experiments (Fig. 2). A laminar airflow was achieved by means of honeycombs network and screens. The air condition unit is powered by a small variable speed electric fan and four controllable heating elements provide a stable air temperature. The air flow velocity was measured by a pitot tube and a pressure transducer during the calibration and test. The output voltage from the hot-wire, pressure transducer output voltage, and the thermometer (NTC) output voltage are transferred to a computer, via an A/D card, having a 12 bit resolution and up to 100 kHz frequency.

The sensing element in our case is a standard 5μm diameter tungsten wire that is heated by an electric current to a temperature of approximately 250 °C. The active wire length is 1.25 mm. For such probes, the convective heat transfer is about 85 percent of the total heat transfer from the heated-wire element [2].

Before measurements, the hot-wire sensor was calibrated in a wind tunnel and the response of the anemometer bridge voltage was also expanded as a least square fit with a 5th order polynomial ($U=C_0+C_1E+C_2E^2+C_3E^3+C_4E^4+C_5E^5$).

The experiments were carried out on a hot wire sensor operating at an over-heat ratio (R_w/R_a) of 1.8. The sensor, after calibration, was tested at different temperatures. The velocity range was 1-2 m/s, which corresponds to a Reynolds number of 0.18-0.35, and the temperature range, was 17.5-40 °C.

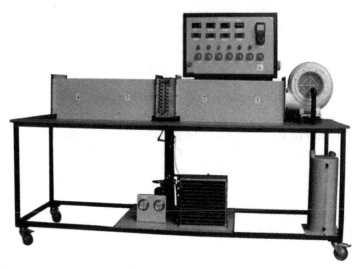

Figure 2. The laboratory air condition unit

4. Results and discussions

To examine the behavior of the hot wire sensor in different conditions and determine the temperature distribution along it, the general hot wire equation must be derived initially. By considering an incremental element of the hot wire, Fig.3, an energy balance can be performed where assume that there is the uniform temperature over its cross-section according to the equation (14).

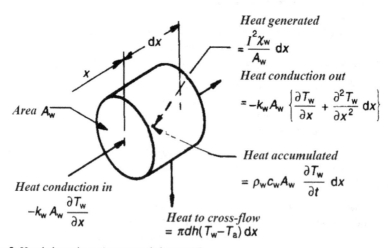

Figure 3. Heat balance for an incremental element [2].

$$\frac{I^2 \chi_w}{A_w} = \pi dh \left(T_w - T_a \right) - K_w A_w \frac{\partial^2 T_w}{\partial x^2} + \pi d\sigma\varepsilon \left(T_w^4 - T_s^4 \right) + \rho_w c_w A_w \frac{\partial T_w}{\partial t} \qquad (14)$$

Where I is electrical current, χ_w is the electrical resistant of the wire material at the local wire temperature, T_w, and A_w is the cross-sectional area of wire where h is the heat-transfer coefficient, c_w is the specific heat of the wire material per unit mass, k_w is the thermal conductivity of the wire material and d is the diameter of wire. With using the fourth-order Runge-Kutta method, this nonlinear secondary differential equation is solved in two conditions: with radiation term and without radiation term. Fig.4 shows the results for this step. As it is shown, the radiation term does not have any effect on the temperature distribution. The previous results achieved in Ref.[3] indicate that, the error due to radiation is in the range 0.1-0.01% and is quite negligible.

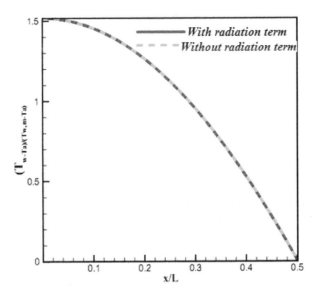

Figure 4. The solution of equation (14) with and without radiation term

Under steady conditions,

$$\frac{\partial T_w}{\partial t} = 0$$

χ_w can be expressed as $\chi_w = \chi_a + \chi_0 \alpha_0 (T_w - T_a)$. Where χ_a and χ_0 are the values of the resistivity at the ambient fluid temperature, T_a, and at 0°C and α_0 is temperature coefficient of resistivity at 0°C. Thus, equation (14) can be rewritten as equation 15 [2]:

$$K_w A_w \frac{d^2 T_w}{dx^2} + \left(\frac{I^2 \chi_o \alpha_o}{A_w} - \pi dh \right) \left(T_w - T_a \right) + \left(\frac{I^2 \chi_a}{A_w} \right) = 0 \qquad (15)$$

With assuming the ambient temperature is constant along the wire, this equation is of the following form (16):

$$\frac{d^2 T_w}{dx^2} + K_1 T_1 + K_2 = 0 \qquad (16)$$

Where

$$T_1 = T_w - T_a$$

And

$$K_1 = \left(\frac{I^2 \chi_o \alpha_o}{A_w} - \pi dh \right)$$

$$K_2 = \left(\frac{I^2 \chi_a}{A_w} \right)$$

The value of K_1 may be negative or positive. Therefore, the solution of equation (16) and temperature distribution along the wire are dependent on the value of K_1. Equation (16) is solved in three states: $K_1 < 0$, $K_1 = 0$, $K_1 > 0$.

4.1. State I: $K_1 < 0$

For more hot-wire applications, K_1 will be negative [2]. In Ref. [2], it is declared that in this state, the solution for a wire of length L will become:

$$T_w = \frac{K_2}{|K_1|} \left[1 - \frac{\cosh\left(|K_1|^{0.5} x \right)}{\cosh\left(\frac{|K_1|^{0.5} L}{2} \right)} \right] + T_a \qquad (17)$$

The mean wire temperature, $T_{w,m}$ is obtained by integrating equation (17):

$$T_{w,m} = \frac{1}{L} \int_{-L/2}^{L/2} T_w(x) dx \qquad (18)$$

Inserting equation (17) into equation (18) gives:

$$T_{w,m} = \frac{K_2}{|K_1|}\left[1 - \frac{\tanh\left(|K_1|^{0.5}L/2\right)}{|K_1|^{0.5}L/2}\right] + T_a \tag{19}$$

The non-dimensional steady state wire temperature distribution will be achieved such as equation (20) [2]:

$$\frac{T_w - T_a}{T_{w,m} - T_a} = \frac{\left[\frac{L\times|K_1|^{0.5}}{2}\right]\left[\cosh\left(\frac{L\times|K_1|^{0.5}}{2}\right) - \cosh\left(x\times|K_1|^{0.5}\right)\right]}{\left[\frac{L\times|K_1|^{0.5}}{2}\right]\cosh\left(\frac{L\times|K_1|^{0.5}}{2}\right) - \sinh\left(\frac{L\times|K_1|^{0.5}}{2}\right)} \tag{20}$$

Where T_a is the ambient fluid temperature and $T_{w,m}$ is the mean wire temperature.

The convective and conductive heat transfer rate can be found from the flow conditions and the wire temperature distribution will earn according to the following equations:

$$\dot{Q}_{cond} = 2k_w A_w \left|\frac{dT_w}{dx}\right|_{x=l/2} \tag{21}$$

$$\dot{Q}_{conv} = \pi dhL(T_{w,m} - T_a) \tag{22}$$

To achieve a reasonable accuracy, the ratio of conductive heat transfer to convective heat transfer should be as low as possible.

$$\frac{\dot{Q}_{cond}}{\dot{Q}_{conv}} = 2\frac{k_w A_w |K_1|^{0.5}}{\pi dLh} \times \frac{\tanh\left(0.5L|K_1|^{0.5}\right)}{\left(1 - \frac{2}{L|K_1|^{0.5}}\tanh\left(0.5L|K_1|^{0.5}\right)\right)} \tag{23}$$

According to equation (23), to reduce the effect of the conductive heat transfer rate, the wire should be as long as possible and the thermal conductive of the wire material should have a low value.

The temperature distribution in the form $(T_w-T_a)/(T_{w,m}-T_a)$, is shown in Fig.5. It is shown that the uniformity of the temperature distribution along the wire increases for longer length wires. Also, the value of temperature in different parts of the wire approaches to mean temperature with escalating the length wire.

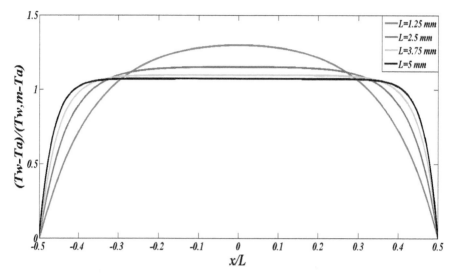

Figure 5. The temperature distribution along a hot wire for various values of L ($K_1 < 0$).

The effect of wire length on the percent of conduction and convection heat transfer is shown in Fig. 6. (Diameter of wire is 5μm and the air velocity is equal to 20 m/s). As it is shown the conductive end losses reduces with increasing the wire length but it should be noted, the maximum value of tanh ($0.5L \mid K_1 \mid^{0.5}$) is approximately 1, so exceeding the wire length over $5.3 / \mid K_1 \mid^{0.5}$ will not cause a reduction in the conductive heat transfer rate.

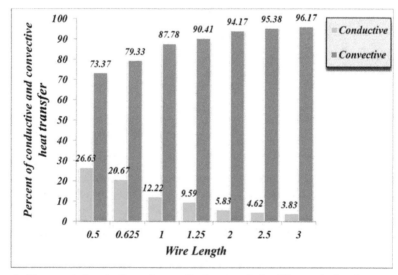

Figure 6. The percent of conduction and convection heat transfer for different wire length.

For hot-wire anemometer applications it is usually advantageous to minimize the rate of conductive heat transfer rate relative to the forced convective heat transfer rate [2]. Fig.7 shows the effect of wire diameter on the non-dimensional temperature distribution. It is shown, the uniformity of temperature distribution decreases with increasing the wire diameter. This variation is due to increasing the wire diameter that will cause the conductive heat transfer rate to the two prongs to be increased as well.

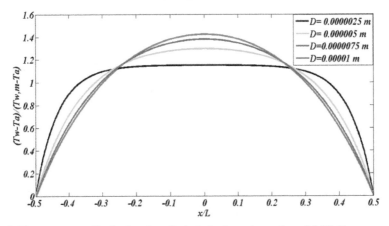

Figure 7. The temperature distribution along the hot wire for various values of d (K1<0).

Comparing between Figs. 6 and 7 shows that the wire diameter has the greater influence on the temperature distribution rather than the wire length.

4.2. State II; K₁=0

In this state, the temperature distribution equation is obtained as:

$$T_w(x) = -\frac{1}{2}K_2L^2\left[\left(\frac{x}{L}\right)^2 - \frac{1}{4}\right] + T_a \tag{24}$$

According to equations (24) and (18), the mean wire temperature is determined as:

$$T_{w,m} = \frac{1}{12}K_2L^2 + T_a \tag{25}$$

By using equations (24) and (25), the non-dimensional wire temperature distribution can be expressed as:

$$\frac{T_w(x) - T_a}{T_m(x) - T_a} = -6\left[\left(\frac{x}{L}\right)^2 - \frac{1}{4}\right] \tag{26}$$

For this state, the non-dimensional temperature distribution is shown in Fig.8. It can be observed that the temperature distribution along the wire is independent of the wire length and for various values of L, all temperature profiles are identical. Also, it can be observed from equation (26) that the non-dimensional temperature distribution does not depend on the wire diameter.

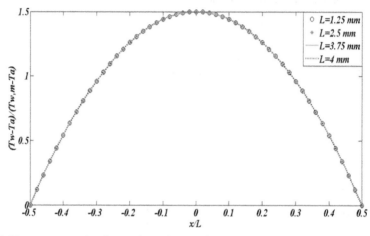

Figure 8. The temperature distribution along a hot wire for various values of L (K1=0).

Furthermore, the ratio of conductive heat transfer rate to the forced convective heat transfer rate can be expressed as:

$$\frac{\dot{Q}_{cond}}{\dot{Q}_{conv}} = \frac{3}{2}\frac{k_w d}{hL} \tag{27}$$

While it is shown, this ratio is directly proportional to the length and it will increase with inverse proportion to the wire diameter.

4.3. State III; K₁>0

Using the mathematical analysis, it can be demonstrated that temperature distribution equation is:

$$T_w = \frac{K_2}{K_1}\left[\frac{\cos\left((K_1)^{0.5}x\right)}{\cos\left(\frac{(K_1)^{0.5}L}{2}\right)} - 1\right] + T_a \tag{28}$$

The mean wire temperature and the non-dimensional wire temperature distribution can be expressed as:

$$T_{w,m} = \frac{K_2}{K_1}\left[\frac{\tan\left(\left(K_1\right)^{0.5} L/2\right)}{\left(K_1\right)^{0.5} L/2} - 1\right] + T_a \tag{29}$$

$$\frac{T_w - T_a}{T_{w,m} - T_a} = \frac{\left[\frac{L\times\left(K_1\right)^{0.5}}{2}\right]\left[\cos\left(x\times\left(K_1\right)^{0.5}\right) - \cos\left(\frac{L\times\left(K_1\right)^{0.5}}{2}\right)\right]}{\sin\left(\frac{L\times\left(K_1\right)^{0.5}}{2}\right)\left[\frac{L\times\left(K_1\right)^{0.5}}{2}\right] - \left[\frac{L\times\left(K_1\right)^{0.5}}{2}\right]\cos\left(\frac{L\times|K_1|^{0.5}}{2}\right)} \tag{30}$$

Fig.9 shows the non-dimensional temperature distribution ($K_1>0$). As it is shown, in this state some fluctuations appear in the temperature profiles. It can be demonstrated from equation (28) that, with approaching the wire length to $\pi/|k_1|^{0.5}$, these fluctuations increases wherever the amplitude oscillatin decreases with growing the wire length.

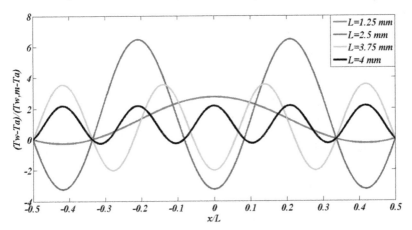

Figure 9. The temperature distribution along a hot wire for various values of L ($K1>0$).

It should be noted that the temperature profile is strongly dependent on the value of K_1 which is relevant to the heat transfer coefficient (Nusselt number). With setting the value of K_1 to zero, one could determine the critical Nusselt number as the following equation:

$$Nu_{critical} = \frac{I^2 \chi_o \alpha_o}{A_w \pi k} \tag{31}$$

In summary, the authors consider the temperature distribution along the hot-wire in the following cases:

Case I, Nu>Nu critical: in this case, increasing the wire length and decreasing the wire diameter will cause the uniformity of temperature distribution to be increased considerably.

Case II, Nu=Nu critical:in this case, the temperature distribution is independent of length and diameter of wire.

Case III, Nu<Nu critical: here, temperature distribution is non-uniform and there are some fluctuations in temperature distribution.

According to equation (4), by knowing E_w, R_w, T_w and T_a in the anemometer, one can calculate the Nu number. The electrical resistance of the wire's material increases linearly with temperature, so that the resistance can be described as:

$$R_w = R_o\left[1 + \alpha_o(T_w - T_o)\right]$$

(32)

Where R_0 is the value of the resistance at a reference temperature T_0 and α is the temperature coefficient of resistance. The recommended value for over-heat ratio is equal to 1.8 and the wire temperature of the chosen probe is then 249.22 °C.

In practical application, the hot-wire anemometer output is bridge voltage E (Fig. 1) whereas for determining the Nusselt number, the value of E_w is required. For a balanced anemometer bridge, the relationship between E (bridge voltage) and E_w (hot-wire sensor voltage) is:

$$E_w = \frac{E}{R_1 + R_w} R_w$$

(33)

For comparison, the calculated Nusselt number by equation (4) that it is based on the fluid properties evaluated at the film temperature defined as the mean of the upstream flow temperature and temperature on the hot wire versus Reynolds number where based on the fluid properties evaluated at the film temperature is presented in Fig.10 with Kramer's experimental formula.

As it is shown, the data does not collapse to one curve and the deviation increases with increasing Reynolds number. Our results are lower than those given by Kramer's formula and the differences may be caused by the effect of conductive heat transfer to the prongs and three-dimensional effect encountered in experiments. In Ref. [2], it is stated that for a standard probe (d=5μm and l=1.25 mm), the conductive heat transfer to the two prongs is about 15 percent of the total heat transfer from the heated-wire element. Although, the results in this study show that at the high velocity, the percentage of error between the predicted Nusselt number by Kramer's formula and the achieved Nusselt number in this study is 50%. This result confirms that, there is a significant difference between the heat transfer process from finite length hot-wire sensor and infinite length one.

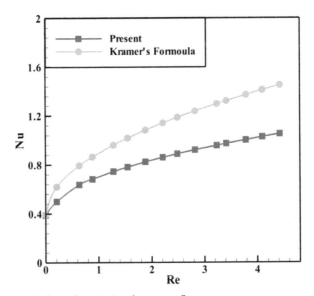

Figure 10. Heat transfer from a hot wire in a free stream flow.

In the next phase, the thermal response of the hot-wire anemometer relative to the velocity and also the air ambient temperature variation is investigated [11]. Figs. 11 and 12 show the variation of convection heat transfer coefficient (h) and Nusselt number versus Reynolds number at different ambient temperatures where fluid properties evaluated at the film temperature $T_f = (T_a + T_w)/2$.

The achieved results indicate convection heat transfer coefficient and Nusselt number vary with variation of air flow temperature and as expected, both of them decrease with increasing the ambient temperature.

Different equations have been proposed to modify the Nusselt number. Lundström et al. [3] claim that it was necessary to evaluate the fluid properties at the air temperature and their results show that evaluating the properties at the film temperature is not enough to achieve a temperature independent calibration law. Collis and Williams [10] realized, using the film reference temperature, that it was necessary to include a temperature loading factor in the Nusselt number King's law according to the following equation;

$$Nu \left(\frac{T_f}{T_a} \right)^{-0.17} = A + BRe^{0.45} \qquad (34)$$

Nusselt number (hd/k) includes both the heat-transfer coefficient and the thermal conductivity of the fluid and these parameters are dependent on the ambient temperature. The temperature role on k can be expressed as:

$$\frac{k}{k_r} = \left(\frac{T}{T_r}\right)^a \tag{35}$$

T and T_r are in absolute temperature but the variation of h with ambient temperature is unknown.

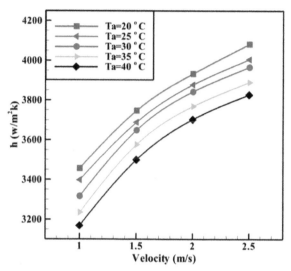

Figure 11. Variations of heat-transfer coefficient Vs. velocity at different ambient temperatures. [11]

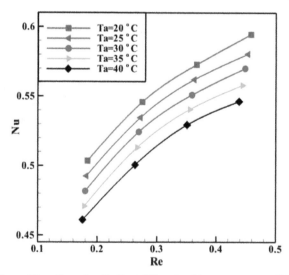

Figure 12. Variations of Nusselt number Vs. Re at different ambient temperatures. [11]

However, the fundamental mechanism for variation of Nusselt number is not known yet but it can be compensated empirically by introducing the modified Nusselt number according to the following equation [11]:

$$Nu_{corr} = Nu\left(\frac{T_a}{T_{a,r}}\right) \tag{36}$$

Where temperatures are in absolute temperature and $T_{a,r}$ is the reference temperature at which the sensor calibration is performed. When the correction is applied to the data in Fig. 12, the data collapse approximately to a single curve as shown in Fig.13.

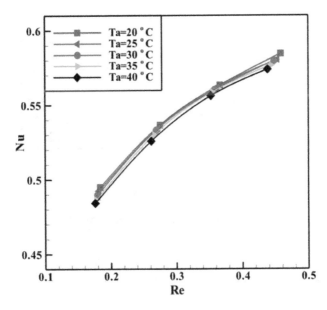

Figure 13. The achieved Nusselt number Vs. Re by employing equation (36) [11].

Fig. 14 presents the response of the CTA on wind speed approximately equal to 1 and 2 m/s at various temperatures. The temperature varies between 22.5 and 37.5°C. As it is shown, the bridge voltage decreases as the higher ambient temperature.

Calibration equations do not include ambient temperature variations, so a correction procedure should be applied. There are three main practical ways [2]:

- Automatic compensation: Use a temperature sensor in the Wheatstone bridge.
- Manual adjustment: Manual adjustment can be made by changing the value of the resistant, R_w, to compensate the changes in T_a.

- Analytical correction: Measure the flow temperature separately and compensate using the heat transfer equation.

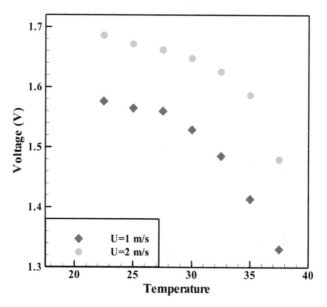

Figure 14. Response of CTA at various ambient temperatures. [11]

In this research, the voltage error due to changes in the ambient temperature is corrected using equation (10) and (37):

$$E_w = E\left(\frac{T_w - T_a}{T_w - T_{a,r}}\right)^{0.5} \times \left(\frac{T_a}{T_{a,r}}\right)^{0.5} \tag{37}$$

Equation (10) only considers the effect of ambient temperature variation but equation (37) regards the effect of Nusselt variation as well the ambient temperature variation.

The percentage error is presented in Fig.15 as a function of flow temperature. The other parameters are reference temperature, $T_{a,r}$=25 °C, average sensor temperature, T_w=249.22 °C and flow velocity, U=1.5 m/s.

It can be seen that the achieved results from equation (37) are more reasonable. It should be noted that by increasing the air temperature, the fluid properties will be changed and these changes have to be taken into consideration. This factor is considered in equation (37) so that at high air temperature it can compensate the ambient temperature variations adequately.

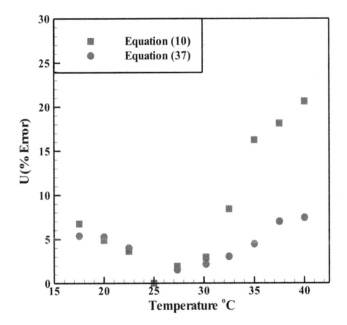

Figure 15. Error in the measurement of flow velocity for the hot wire sensor [11]

5. Conclusion

The analytical solutions for heat transfer equation of hotwire indicate that the temperature distribution along a hot-wire sensor is dependent on the critical Nusselt number. Nu<Nu$_{critical}$ leads to increasing the wire length and decreasing the wire diameter which will cause the uniformity of temperature distribution to be increased considerably. For Nu=Nu$_{critical}$, wire length and diameter don't have any effect on the temperature distribution. If Nu>Nu$_{critical}$, some fluctuations will appear in the temperature profiles.

The results from experimental investigation show that the values of both the air temperature and Nusselt number have influence on the output voltage of the CTA. In this study, two ways have been employed to compensate the ambient temperature changes. In the first case, the effect of ambient temperature variation is only considered but in the second case, the effect of Nusselt variation is also regarded. At low temperature variation, the accuracy of two methods is almost the same whereas by increasing the air temperature, the second method which consider the changes in fluid properties, provide more accurate results in compare with the first method.

6. Appendix

Aw   Surface area of sensor, m^2

Cw   Specific heat of the wire material per unit mass, J/Kg. K

D   Wire diameter, m

E   Bridge voltage ,V

E_{corr} Corrected voltage, V

E_w Hot-wire sensor voltage, V

H   heat-transfer coefficient, $W/m^2.K$

I   Electrical current passing through the sensor, A

K   Thermal conductivity of the fluid, W/m. K

k_w Thermal conductivity of the wire material, W/m. K

L   Wire length, m

M   Temperature loading factor used in equation (10)

N   Constant used in equation (7)

Nu   Nusselt Number, hd/k

Pr   Prandtl number, $\mu Cp/K$

$\dot{Q}_{elec}$ Electrical power dissipation, W

Re   Reynolds number, ud/v

R_w Sensor resistance

T_a Ambient temperature during data acquisition, K

$T_{a,r}$ Ambient reference temperature during sensor calibration, K

T_w Temperature of hot-wire sensor, K

$T_{w,m}$ Mean wire temperature, K

U   Instantaneous velocity, m/s

A_o Temperature coefficient of resistivity at 0°C

E   Emissivity of the sensor

Σ Stefan-Boltzmann constant

χ_o Values of resistivity at 0°C

χ_a Values of resistivity at the ambient fluid temperature

M   Dynamic viscosity, N.s/m²

Author details

Mojtaba Dehghan Manshadi and Mohammad Kazemi Esfeh
Malekashtar University of Technology, Iran
University of Yazd, Yazd, Iran

7. References

[1] Manshadi M. D. (2011). "The Importance of Turbulence in Assessment of Wind Tunnel Flow Quality", Book chapter No. 12, Wind Tunnels and Experimental Fluid Dynamics Research, Edited by Jorge Colman Lerner and Ulfilas Boldes, Intech publisher.

[2] Bruun H H. Hot-wire Anemometry, Principles and Signal Analysis. Oxford University Press; 1995.

[3] LundstrÖm H, Sandberg M, Mosfegh B. Temperature dependence of convective heat transfer from fine wires in air: A comprehensive experimental investigation with application to temperature compensation in hot-wire anemometry. Experimental Thermal and Fluid Science 2007; 32: 649-657.

[4] Kanevce G, Oka S. Correcting hot-wire readings for influence of fluid temperature variations 1973. DISA Info, No. 15, 21-24.

[5] Benjamin SF, Roberts CA. Measuring flow velocity at elevated temperature with a hot wire anemometer calibrated in cold flow. Heat and Mass Transfer 2002; 45: 703-706.

[6] Improved temperature correction in stream Ware®. DANTEC DYNAMICS. Technical Note of dynamics, Publication No.TN049909, P2, 2002.

[7] Ardakani MA, Farhani F. Experimental study on response of hot wire and cylindrical hot film anemometers operating under varying fluid temperatures. Flow Measurement and Instrumentation 2009; 20: 174_179.

[8] Suminska OA. Application of a constant temperature anemometer for balloon-borne stratospheric turbulence soundings. MSc thesis. University of Rostock; 2008.

[9] Kramers H. Heat transfer from spheres to flowing media. Physica 1946;12:61-80.

[10] Collis DC, Williams MJ. Two-dimensional convection from heated wires at low Reynolds numbers. J. Fluid Mech 1959; 6:357-384.

[11] Dehghan Manshaid M, Kazemi Esfeh M. A new approach about heat transfer of hot-wire anemometer. Accepted to publishing in the Applied Mechanics and Materials Journal, 2012.

Chapter 3

Experimental Determination of Heat Transfer Coefficients During Squeeze Casting of Aluminium

Jacob O. Aweda and Michael B. Adeyemi

Additional information is available at the end of the chapter

1. Introduction

Casting process is desired because it is very versatile, flexible, and economical and happens to be the shortest and fastest way to transform raw material into finished product. Squeeze casting belongs to permanent mould casting method which offers considerable saving in cost for large production quantities when the size of the casting is not large. Squeeze casting has the advantage of producing good surface finish, close dimensional tolerance and the absence of sand inclusions on the cast surfaces of the products as opined by Das and Chatterjee, (1981).

The solidification process of the molten aluminium metal in the steel mould takes a complex form, (Hosford and Caddell, 1993) and (Potter and Easterling, 1993). During solidification all mechanisms of heat transfer are involved and the solidifying metal undergoes state and phase changes. The final structure and properties of the cast product obtained depend on the casting parameters applied i.e. applied pressures, die pre-heat temperature, delay time and period of applied pressure on the solidifying metal, (Potter and Easterling, 1993), (Bolton, 1989) and (Callister, 1997). The prediction of temperature distribution and solidification rate in metal casting is very important in modern foundry technologies. This helps to control the fundamental parameters such as the occurrence of defects, as well as, the influence on final properties of cast products and the mould wall / cast metal interface contact surface.

Heat transfer coefficients during squeeze cast of commercial aluminium were determined using the solidification temperature versus time curves obtained for varying applied pressures during squeeze casting process. The steel mould / cast aluminium metal interface temperatures versus times curve obtained through polynomial curves fitting and extrapolation was compared with the numerically obtained temperatures versus times curve. Interfacial heat transfer coefficients were determined experimentally from measured values of heating and cooling temperatures of steel mould and cast metal and compared with the numerically obtained values and found to be fairly close in values.

Aluminium is a product with unique properties, making it a natural partner for the building and other manufacturing industries. The commercially pure aluminium metal used for this research work finds extensive use in the building, manufacturing and process industries, both as a material of construction and household goods. Products of squeeze casting are of improved mechanical properties and could be given heat treatment. Heat dissipation from the squeeze cast specimen is fast thus producing products of fine grains as compared to the slow cooling of sand casting, which produces large grains. Products obtained through squeeze casting are with improved mechanical properties.

2. Squeeze casting procedure

A metered quantity of molten metal was poured into the steel mould cavity at a supper-heat temperature of between 40-60 °C fast but avoiding turbulence. The upper die was then released to close the mould cavity with and without applying any load on the upper die. Thermocouples were inserted into the drilled holes made in the die, which were used to monitor both the die and cast metal temperatures. The terminals of the thermocouples were connected to the chart recorder/plotter (set at the highest speed of 10mm/s and voltage 100mV) through the cold junction apparatus, maintained at 0 °C throughout the measuring period.

3. Assumptions made

i.   Heat transfer in the molten metal cast zone is due to both conduction and convection while conduction heat transfer occurs in the steel mould, it is convection at the outer surface of the steel mould.

ii.   The thickness of the cast specimen is much smaller than the diameter (radial dimension), thus giving one dimensional heat transfer process.

iii. Considering the symmetrical nature of the cast specimen, solidification process was assumed symmetrical and only lower half of the specimen's thickness was analysed see figure 1.

iv.   The bottom of the squeeze casting rig was and the heat losses to the atmosphere was small and neglected.

v.   Heat losses through conduction and convection to the atmosphere at the punch were neglected, as a result of short time of pressure application.

vi.   The process of analyses in the cast specimen starts only when the steel mould cavity had been filled with the required quantity of liquid molten metal (i.e. heat transfer processes during pouring of molten aluminium into the steel mould are not considered).

vii. Density of the molten and solidified aluminium metal was assumed to be the same and independent of temperature.

viii. Thermal conductivity and specific heat of aluminium metal were dependent on the cast temperatures.

4. Heat transfer governing equations

4.1. Without pressure application on the cast metal

A measured quantity of molten aluminium metal was poured into the steel mould cavity. The process of solidification begins from the steel mould/cast metal interface and continues inwards into the cast metal. As this process continues, there was an increase in the thickness of the solidified layer and a decrease in the liquid molten metal portion. For the situation when no pressure was applied on the solidified molten metal, the governing heat transfer equations in one dimension are given by equation (1).

$$\rho \, C \frac{\partial T}{\partial t} = K \left[\frac{\partial^2 T}{\partial r^2} + \frac{1}{r} \frac{\partial T}{\partial r} \right] \tag{1}$$

From figure 1, equation (1) is defined within the region with;

i.   Steel mould,

$$L \le r_{st} \le (L + Q)$$

ii.   Solidified molten,

$$\left(L - X_r^j \right) \le r_S \le L$$

$$T_S = T_{MM} = 660 \left({}^0C \right) \tag{2}$$

iii.   Liquid molten metal,

$$0 \le r_L \le \left(L - X_r^j \right)$$

$$K_S \frac{\partial T_S}{\partial r} = 0 \tag{3}$$

$$r = 0 \tag{4}$$

$$T_L = T_P = 720 \left({}^0C \right) \tag{5}$$

iv.   At the phase change boundary condition;

$$\rho_L L_f \frac{dX_r^j}{dt} = K_L \frac{\partial T_L}{\partial r} - K_S \frac{\partial T_S}{\partial r} \tag{6}$$

where,

$$r = L - X_r^j \tag{7}$$

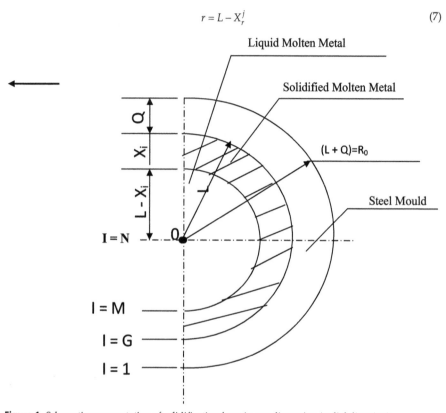

Figure 1. Schematic representation of solidification front in one dimension (radial direction)

4.2. Casting with pressure application on the solidified molten metal

As the cast aluminium metal solidifies, pressure is applied on the specimen, observing lapse or delay time, t_l, while varying the values of pressure applied. The time between the end of pouring of molten metal and pressure application known as lapse time, is recorded. This is necessary such that the cast specimen will not stick to the upper punch or cause the cast metal to tear with pressure application. Due to the applied pressure, an internal energy Δq is generated within the solidified molten metal, (see figure 2).

Inserting the internal energy into the heat transfer equation (1) for solidified molten metal, it becomes equation (8),

$$\rho_S C_S \frac{\partial T_S}{\partial t} = K_S \left[\frac{\partial^2 T_S}{\partial r^2} + \frac{1}{r_S} \frac{\partial T_S}{\partial r} \right] + \Delta q \tag{8}$$

where,

$$\nabla q = \nabla q_P + \nabla q_f \tag{9}$$

Δq -internal energy generated by applied pressure,
Δq_P -energy due to plastic strain within the solidified molten metal material,
Δq_f -frictional energy generated during pressure application,

$$\nabla q_f = \nabla q_{fP} + \nabla q_{fm} \tag{10}$$

Δq_{fP} -frictional energy due to punch / solidified molten metal interface,
Δq_{fm} -frictional energy due to steel mould cylindrical surface / solidified molten metal interface.

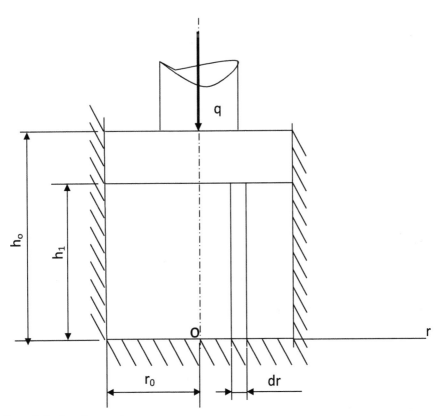

Figure 2. Cast specimen under pressure

5. Formulation of heat transfer equations

Finite difference expressions for the nodal temperatures are obtained by either energy balance within an elemental volume around the node or by substitutions into the governing partial differential equations. Partial time derivative of temperature contained in the equations can be written in terms of a moving gradient as in White (1991) at a velocity of dx/dt. In the heat transfer equation, temperature is defined as a function of distance, r with time, t and represented in equation (11).

$$T = T(r, t) \tag{11}$$

5.1. In the steel mould

$$\frac{dT_{st}}{dt} = \alpha_{st} \frac{\partial^2 T_{st}}{\partial r^2} + \frac{\alpha_{st}}{r_{st}} \frac{\partial T_{st}}{\partial r} \tag{12}$$

where,

$$\frac{dr_{st}}{dt} = 0 \tag{13}$$

$$\alpha_{st} = \frac{K_{st}}{\rho_{st} C_{st}} \tag{14}$$

α_{st} -thermal diffusivity of steel mould material.

5.2. In the solidified molten metal

$$\frac{dT_S}{dt} - \frac{(I-G)}{(M-G)} \frac{dX_i^r}{dt} \frac{\partial T_S}{\partial r} = \alpha_S \frac{\partial^2 T_S}{\partial r^2} + \frac{\alpha_S}{r_S} \frac{\partial T_S}{\partial r} \tag{15}$$

where,

$$\alpha_S = \frac{K_S}{\rho_S C_S} \tag{16}$$

α_S -thermal diffusivity of solidified molten metal material.

5.3. In the liquid molten metal

$$\frac{dT_L}{dt} - \frac{(I-M)}{(N-M)} \frac{dX_i^r}{dt} \frac{\partial T_L}{\partial r} = \alpha_L \frac{\partial^2 T_L}{\partial r^2} + \frac{\alpha_L}{r_L} \frac{\partial T_L}{\partial r} \tag{17}$$

where,

$$\alpha_L = \frac{K_L}{\rho_L C_L} \tag{18}$$

α_L -thermal diffusivity of liquid molten metal material.

6. Nodal divisions

The steel mould, the solidified metal and the molten metal regions were discretized separately. Each of these regions was divided into a fixed number of gridal points as in figure 1.

6.1. In the steel mould

$$d_{st} = \frac{Q}{(G-I)} \tag{19}$$

I = 1, 2, 3,..., G-1

6.2. In the solidified molten metal

$$d_S = \frac{X_r^i}{(M-I)} \tag{20}$$

I = G+1, G+2, G+3, ... , M-1

6.3. In the liquid molten metal portion

$$d_L = \frac{\left(L - X_r^i\right)}{(N-I)} \tag{21}$$

I = M+1, M+2, M+3 ... N-1

6.4. In the phase change boundaries

The phase change is represented by the equations;

$$d_{Ps} = \frac{X_r^i}{(M-G)} \tag{22}$$

$$d_{PL} = \frac{\left(L - X_r^i\right)}{(N-M)} \tag{23}$$

6.5. At the completion of solidification

$$d_{SC} = \frac{L}{(N-G)} \tag{24}$$

As solidification time progresses, the boundary locations change, and the thickness of the solidified molten metal in the radial direction increases. The rate of change of boundary location with time is represented, mathematically in equation (25),

$$\frac{dX_r^i}{dt} = \frac{\left(X_r^{j+1} - X_r^j\right)}{\delta} \tag{25}$$

Where,

$\left(X_r^{j+1} - X_r^j\right)$ -are differences in the thickness of solidified molten metal at a particular time interval as time progresses in the radial direction.

δ -time interval.

7. Boundary conditions

The problem of phase change during solidification is that the location of the solidifying molten metal / liquid molten metal interface is not known and this is determined continuously by appropriate mathematical analysis. This moving interface is normally expressed mathematically by the energy balance equations at the interfaces. In the numerical analysis, as solidification of molten aluminium metal progresses, three boundary interfaces occurred as:

7.1. Steel mould-atmosphere interface (I = 1)

The heat conducted to the steel mould material (from I = 2 to 1 = 1) equals the sum of the change in the internal energy and heat convected from the surface of the steel metal mould material into the atmosphere, mathematically represented in equation (26).

$$\frac{\partial T_{st}}{\partial t} = \frac{2K_{st}}{\rho_{st}C_{st}d_{st}} \frac{\partial T_{st}}{\partial r} - \frac{2H^*}{\rho_{st}C_{st}d_{st}}(T_i^j - T_\infty) \tag{26}$$

I = 1

7.2. In the solidified molten metal-steel mould interface (I = G)

The sum of the heat conducted from the solidifying molten metal/steel mould interface and the change (decrease) in the internal energy at the boundary equal the sum of the heat conducted to the steel mould and the change (increase) in the internal energy at the interface.

$$K_S \frac{\partial T_S}{\partial r} + \frac{1}{2}\rho_S C_S d_S \frac{\partial T_S}{\partial t} = K_{st} \frac{\partial T_{st}}{\partial r} + \frac{1}{2}\rho_{st} C_{st} d_{st} \frac{\partial T_{st}}{\partial t} \qquad (27)$$

$I = G$

7.3. In the liquid molten metal – Solidified molten metal interface, (I = M)

The sum of the heat conducted from the liquid molten metal and the internal energy generated equal the sum of heat conducted to the solidified molten metal and the internal energy generated at the interface as in equation (28).

$$K_L \frac{\partial T_L}{\partial r} - \frac{1}{2}\rho_L C_L d_L \frac{\partial T_L}{\partial t} = K_S \frac{\partial T_S}{\partial r} + \frac{1}{2}\rho_S C_S d_S \frac{\partial T_S}{\partial t} \qquad (28)$$

$I = M$

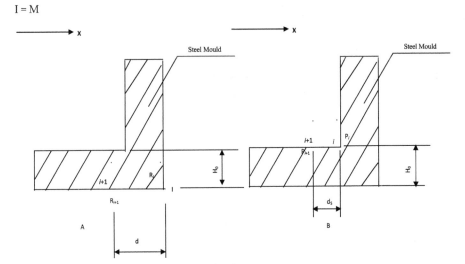

Figure 3. Corner nodes. A) External, B) Internal nodes

7.4. At the corners

The usual one dimensional heat transfer analysis does not take into consideration heat loss at the corners as represented in figure 3. Heat loss at both the external and internal corner nodes of the steel mould have been considered and analysed while the die is lagged at the bottom surface.

7.5. External corner effect

$$q_{i+1} + q_g = q_{conv} \qquad (29)$$

The heat conducted from point R_{i+1} to point R_i added to the change in the internal energy equal the amount of heat convected out to the atmosphere at point R_L, Fig. (3a). The finite form of equation (29) is represented by equation (30),

$$\frac{K_{st}}{d_{st}}\frac{\partial T_{st}}{\partial r} - \frac{1}{2}\rho_{st}C_{st}\frac{\partial T_{st}}{\partial t} = \frac{h}{d_{st}}\left(T_i^j - T_\infty^j\right) + \frac{h}{H}\left(T_i^j - T_\infty^j\right) \tag{30}$$

7.6. Internal corner effect

The heat conducted from point P_{i+1} to point P_I added to the change in the internal energy amount to the heat conducted into the steel mould at point I, fig. (3b) thus becoming equation (31);

$$\frac{K_S}{d_{st}}\frac{\partial T_S}{\partial r} + \frac{1}{2}\rho_S C_S\frac{\partial T_{st}}{\partial t} = \frac{K_{st}}{d_S}\frac{\partial T_{st}}{\partial r} + \frac{K_{st}}{H1}\frac{\partial T_{st}}{\partial Z} \tag{31}$$

7.7. First time analysis

Solidification takes place only in the radial direction, a one-dimensional heat solidification problem was assumed numerically to take place in the radial direction only after filling the steel mould cavity with the liquid molten metal.

For the first time analysis, the specimen is considered to be in the molten stage and therefore equation (32) for the liquid molten metal is used for computation. Thus;

$$\rho_L C_L \frac{dT_L}{dt} = K_L\left[\frac{\partial^2 T_L}{\partial r^2} + \frac{1}{r_L}\frac{\partial T_L}{\partial r}\right] \tag{32}$$

This equation is subjected to the boundary conditions with;

$$T_L = T_S \tag{33}$$

The instantaneous radius r_i in the first time analysis is given by equation (34);

$$r_L = \frac{(N-I)}{(N-G)}L \tag{34}$$

I = G, G+1, G+2, G+3,...., N

The boundary velocity of moving coordinate of equation (32) is given as equation (35);

$$\frac{dr_i}{dt} = \frac{d}{dt}\left[\frac{(N-I)}{(N-G)}L\right] = 0 \tag{35}$$

7.8. At the completion of solidification

At the completion of solidification, the whole molten region becomes solidified molten metal, and just before pressure is applied, the governing heat transfer equation becomes equation (36) representing the solidified portion;

$$\rho_s C_s \frac{\partial T_L}{\partial t} = K_s \left[\frac{\partial^2 T_S}{\partial r^2} + \frac{1}{r_{SC}} \frac{\partial T_S}{\partial r} \right] \tag{36}$$

The equation is applicable within the region defined by equation (37);

$$r = L \tag{37}$$

The boundary motion at the completion of solidification is as in equation (38);

$$r_{SC} = \frac{(N-I)}{(N-G)} L \tag{38}$$

I = G, G+1, G+2, G+3,…, .N

The boundary velocity at the completion of solidification is expressed as equation (39);

$$\frac{dr_{SC}}{dt} = \frac{d}{dt} \left[\frac{(N-I)}{(N-G)} L \right] = 0 \tag{39}$$

where,

L -constant value

8. Finite difference of governing heat transfer equations

The heat transfer equations generated in the cast metal and interfaces are written in the finite difference forms. These equations are presented in the various regions thus;

8.1. In the steel mould region

$$T_i^{j+1} = \left[1 - \frac{2\delta \alpha_{st}}{d_{st}^2} - \frac{\delta \alpha_{st}}{r_{st} d_{st}} \right] T_i^j + \left[\frac{\delta \alpha_{st}}{d_{st}^2} + \frac{\delta \alpha_{st}}{r_{st} d_{st}} \right] T_{i+1}^j + \frac{\delta \alpha_{st}}{d_{st}^2} T_{i-1}^j \tag{40}$$

where,

$$r_{st} = L + \frac{Q(G-I)}{(G-1)} \quad \text{(a)}$$

$$\tag{41}$$

$$d_{st} = \frac{Q}{(G-I)} \quad \text{(b)}$$

I = 1,2,3,……………G

8.2. In the solidified molten metal portion

$$T_i^{j+1} = \left[1 - \frac{(I-G)}{(M-G)}\frac{\left(X_r^{j+1} - X_r^j\right)}{d_S} - \frac{2\delta\,\alpha_S}{d_S^2} - \frac{\delta\,\alpha_S}{r_S d_S}\right]T_i^j +$$

$$+ \left[\frac{(I-G)}{(M-G)}\frac{\left(X_r^{j+1} - X_r^j\right)}{d_S} + \frac{\delta\,\alpha_S}{d_S^2} + \frac{\delta\,\alpha_S}{r_S d_S}\right]T_{i+1}^j + \frac{\delta\,\alpha_S}{d_S^2}T_{i-1}^j$$

(42)

where,

$$r_S = L - \frac{X_r^j(I-G)}{(M-G)} \quad \text{(a)}$$

$$d_S = \frac{X_r^j}{(M-I)} \quad \text{(b)}$$

(43)

I = G+1, G+2, G+3,..........., M

8.3. In the liquid molten metal region

$$T_i^{j+1} = \left[1 - \frac{(I-M)}{(N-M)}\frac{\left(X_r^{j+1} - X_r^j\right)}{d_L} - \frac{2\delta\,\alpha_L}{d_L^2} - \frac{\delta\,\alpha_L}{r_L d_L}\right]T_i^j +$$

$$\left[\frac{(I-M)}{(N-M)}\frac{\left(X_r^{j+1} - X_r^j\right)}{d_L} + \frac{\delta\,\alpha_L}{d_L^2} + \frac{\delta\,\alpha_L}{r_L d_L}\right]T_{i+1}^j + \frac{\delta\,\alpha_L}{d_L^2}T_{i-1}^j$$

(44)

where,

$$r_L = \left(L - X_r^j\right)\frac{(I-M)}{(N-M)} \quad \text{(a)}$$

$$d_L = \frac{\left(L - X_r^j\right)}{(N-I)} \quad \text{(b)}$$

(45)

I = M, M+1, M+2, M+3, ..., N

8.4. In the phase change boundary condition (I = M)

$$X_r^{j+1} = X_r^j - \left[\frac{\delta K_L}{\rho_L L_f h_{pL}} + \frac{\delta K_S}{\rho_L L_f h_{ps}}\right]T_i^j + \frac{\delta K_L}{\rho_L L_f h_{pL}}T_{i+1}^j + \frac{\delta K_S}{\rho_L L_f h_{ps}}T_{i-1}^j$$

(46)

where,

$$h_{PL} = \frac{\left(L - X_r^j\right)}{\left(N - M\right)} \quad \text{(a)}$$

$$h_{Ps} = \frac{X_r^j}{\left(M - G\right)} \quad \text{(b)}$$

(47)

8.5. In the steel mould / atmosphere interface (I = 1)

$$T_i^{j+1} = \left[1 - \frac{2\delta K_{st}}{d_{st}^2 \rho_{st} C_{st}}\right] T_i^j + \frac{2\delta K_{st}}{d_{st}^2 \rho_{st} C_{st}} T_{i+1}^j - \frac{2\delta H^*}{d_{st} \rho_{st} C_{st}} \left(T_i^j - T_\infty^i\right)$$

(48)

I = 1

where;

$$d_{st} = \frac{Q}{(G-1)}$$

(49)

I = 1, 2, 3, ..., G

8.6. In the solidified molten metal / steel mould interface (I = G)

$$T_i^{j+1} = a\left[\frac{2\delta K_S}{d_S} + d_S \rho_S C_S + \rho_S C_S \frac{(I-G)}{(M-G)}\left(X_r^{j+1} - X_r^j\right) + \frac{2\delta K_{st}}{d_{st}} - d_{st} \rho_{st} C_{st}\right] T_i^j$$

$$+ a\left[\rho_S C_S \frac{(I-G)}{(M-G)}\left(X_r^{j+1} - X_r^j\right) - \frac{2\delta K_S}{d_S}\right] T_{i+1}^j - \frac{2a\delta K_{st}}{d_{st}} T_{i-1}^j$$

(50)

where,

$$a = \frac{1}{\left(d_S \rho_S C_S - d_{st} \rho_{st} C_{st}\right)} \quad \text{(a)}$$

$$d_{st} = \frac{Q}{(G-I)} \quad \text{(b)}$$

I = 1, 2, 3,..., G

$$d_S = \frac{X_r^i}{(M-I)} \quad \text{(c)}$$

I = G+1, G+2, G+3,..., M

(51)

8.7. In the liquid molten metal / solidified molten metal interface (I = M)

$$T_i^{j+1} = b \begin{bmatrix} d_L\rho_L C_L - \rho_L C_L \dfrac{(I-M)}{(N-M)}\left(X_r^{j+1} - X_r^j\right) - \dfrac{2\delta K_L}{d_L} + d_S\rho_S C_S - \dfrac{2\delta K_S}{d_S} + \\[2mm] \rho_S C_S \dfrac{(I-G)}{(M-G)}\left(X_r^{j+1} - X_r^j\right) \end{bmatrix} T_i^j$$

$$(52)$$

$$+ b\left[\dfrac{2\delta K_S}{d_S} - \rho_S C_S \dfrac{(I-G)}{(M-G)}\left(X_r^{j+1} - X_r^j\right)\right] T_{i-1}^j$$

$$+ b\left[\dfrac{2\delta K_L}{d_L} + \rho_L C_L \dfrac{(I-M)}{(N-M)}\left(X_r^{j+1} - X_r^j\right)\right] T_{i+1}^j$$

where,

$$b = \dfrac{1}{\left(d_S\rho_S C_S + d_L\rho_L C_L\right)} \quad \text{(a)}$$

$$d_S = \dfrac{X_r^j}{(M-I)} \quad \text{(b)}$$

$$(53)$$

$$d_L = \dfrac{\left(L - X_r^j\right)}{(N-I)} \quad \text{(c)}$$

I = M, M+1, M+2, M+3,...,N

8.8. External corner effect (I = 1)

$$T_i^{j+1} = \left[1 - \dfrac{2\delta K_{st}}{\rho_{st}C_{st}d_{st}^2} - \dfrac{2\delta h}{\rho_{st}C_{st}d_{st}} - \dfrac{2\delta h}{\rho_{st}C_{st}H}\right] T_i^j +$$

$$+ \dfrac{2\delta K_{st}}{\rho_{st}C_{st}d_{st}^2} T_{i+1}^j +$$

$$(54)$$

$$+ \left[\dfrac{2\delta h}{\rho_{st}C_{st}d_{st}} + \dfrac{2\delta h}{\rho_{st}C_{st}H}\right] T_\infty^j$$

where,

$$d_{st} = \dfrac{Q}{(G-I)} \quad (55)$$

8.9. Internal corner effect (I = G)

$$T_i^{j+1} = \left[1 + \frac{2\delta K_S}{\rho_S C_S d_S^2} + \frac{(I-G)}{(M-G)} \frac{\left(X_r^{j+1} - X_r^j\right)}{d_S} + \frac{2\delta K_{st}}{\rho_S C_S d_S d_{st}} + \frac{2\delta K_{st}}{\rho_S C_S H1 d_{st}} \right] T_i^j -$$

(56)

$$-\frac{2\delta K_S}{\rho_S C_S d_S^2} T_{i+1}^j - \left[\frac{2\delta K_{st}}{\rho_S C_S d_S d_{st}} + \frac{2\delta K_{st}}{\rho_S C_S H1 d_{st}} \right] T_{i-1}^j$$

where,

$$d_S = \frac{X_r^j}{(M-I)} \qquad \text{(a)}$$

$$I = G+1, G+2, G+3, \dots, M$$

(57)

$$d_{st} = \frac{Q}{(G-I)} \qquad \text{(b)}$$

$$I = 1, 2, 3, \dots, G$$

8.10. First time analysis

Finite difference form of equation (32) therefore becomes equation (58);

$$T_i^{j+1} = \left[1 - \frac{2\delta\alpha_L}{d_L^2} - \frac{\delta\alpha_L}{r_L d_L} \right] T_i^j + \left[\frac{\delta\alpha_L}{d_L^2} + \frac{\delta\alpha_L}{r_L d_L} \right] T_{i+1}^j + \frac{\delta\alpha_L}{d_L^2} T_{i-1}^j \qquad (58)$$

where,

$$r_L = \frac{(N-I)}{(N-G)} L \qquad \text{(a)}$$

(59)

$$d_L = \frac{L}{(N-G)} \qquad \text{(b)}$$

I = G+1, G+2, G+3, ………., .N

8.11. At the completion of solidification

Finite difference form of equation (36) becomes equation (60);

$$T_i^{j+1} = \left[1 - \frac{2\delta\alpha_S}{d_{SC}^2} - \frac{\delta\alpha_S}{r_{SC} d_{SC}} \right] T_i^j + \left[\frac{\delta\alpha_S}{d_{SC}^2} + \frac{\delta\alpha_S}{r_{SC} d_{SC}} \right] T_{i+1}^j + \frac{\delta\alpha_S}{d_{SC}^2} T_{i-1}^j \qquad (60)$$

where;

$$r_{SC} = \frac{(N-I)}{(N-G)} L \qquad \text{(a)}$$

$$d_{SC} = \frac{L}{(N-G)} \qquad \text{(b)}$$

(61)

$I = G+1, G+2, G+3, \ldots , (N-1)$

9. Stability criteria

For stability criteria to be achieved, the values of temperature T_i^j in all the heat governing equations should not be negative according to Ozisik (1985) and White (1991) not to negate the law of thermodynamics which could lead to temperature fluctuations. Therefore, for stability to be achieved, the coefficients of T_i^j in each of the equations must be greater than zero.

10. Casting with pressure application and die heating

Pressure was applied only when the cast specimen was solidified, the governing heat transfer equation therefore, takes the form of solidified molten metal (completion of solidification). The finite difference of equation (8) is written as equation (62);

$$T_i^{j+1} = \left[1 - \frac{2\delta \alpha_S}{d_S^2} - \frac{\delta \alpha_S}{r_S d_S} \right] T_i^j + \left[\frac{\delta \alpha_S}{d_S^2} + \frac{\delta \alpha_S}{r_S d_S} \right] T_i^j + \frac{\delta \alpha_S}{d_S^2} T_{i-1}^j + \Delta T$$

(62)

where,
ΔT -temperature change resulting from pressure application

The cast specimen height, h_c, is pressure dependent and the relationship is expressed with the equation (63) below after performing series of experiment with various applied pressure;

$$h_c = -0.00007P + 0.036833$$

(63)

where coefficient of correlation r = 0.996

The plastic flow stress, $\sigma_{(T)}$, is dependent on both the applied pressure, P, and die temperature, TM, (White, 1991), and expressed with the equation (64);

$$\sigma(T) = 0.244P - 0.0405TM + 18.614$$

(64)

where coefficient of correlation r = 0.9508

11. Casting with die heating

Aluminium cast specimens were produced with die pre- heating temperatures of between 100- 300⁰C without applying pressure on the solidifying aluminium metal. The die heating process was carried out, using three electric heater rods (100Watts each) that were connected to a.c supply. The required die temperatures were set and controlled, using a bimetallic thermostat.

12. Heat transfer coefficient evaluations

The method of calculating heat transfer coefficients as reported by Santos et al (2001) and Maleki et al, (2006) is based on the knowledge of known temperature histories at the interior points of the casting or mould together with the numerical models of heat flow during solidification. These temperatures are difficult to measure due to the difficulty in locating accurate position of thermocouple at the interface. Therefore, the inverse heat conduction problem based on non-linear estimation technique of Chattopadhyay, (2007) and Hu and Yu, (2002), has been adopted to determine the values of interface heat transfer coefficients, as a function of time during solidification of squeeze casting. Solidification of squeeze casting of aluminium involves phase change and therefore thermal properties of aluminium are temperature dependent, making the inverse heat conduction problem non-linear.

The governing heat transfer equation in one-dimensional cylindrical coordinates is given by equation (65):

$$\rho c_p \frac{\partial T}{\partial t} = \frac{1}{r} \frac{\partial}{\partial r} \left(K_{al} r \frac{\partial T}{\partial r} \right)$$

(65)

Equation (65) holds within the boundary condition as expressed in equation (66):

$$q = K_{al}(T) \frac{\partial T}{\partial r} = h_{al}(T) \left[T_{al} - T_M \right]$$

(66)

The thermal conductivity K_{al} (Reed-Hill and Abbaschian 1973) and (Elliot, 1988) of aluminium is dependent upon casting temperature, T_{al}, and expressed in equation (67):

$$K_{al}(T) = 241.84 - 0.041T_{al}$$

(67)

The heat flow across the casting/mould interface can be characterized by an average interfacial heat transfer coefficient, $h_{al}(T)$ as obtained by Gafur et al (2003) and Santos et al (2004). This is expressed mathematically in equation (68):

$$h_{al}(T) = \frac{q}{\left[T_{al} - T_M \right]}$$

(68)

The heat transfer coefficient, h, at the interface is estimated by minimizing the errors between numerically estimated and measured temperatures defined by equation (69):

$$F(h) = \sum_{i=1}^{n} \left(T_{est} - T_{exp} \right)^2$$

(69)

where,

T_{est} and T_{exp} -are the estimated and experimentally measured temperatures at various thermocouples location and times,

n -iteration stage

13. Numerical simulations of differential equations

Squeeze casting consists of two stages, the first of which is mould filling: - the mould is filled with the required quantity of liquid molten metal; the second is cooling, this continues until the part has solidified completely. Controlling both stages is of major importance for obtaining sound casts with the required geometry and mechanical properties as observed by (Kobryn and Semiatin, (2000), Browne and O'Mahoney, (2001) and Martorano and Capocchi, (2000). When molten metal is poured into the mould cavity, it is initially in the liquid state with a high fluidity. It quickly becomes very viscous, in the early stage of solidification, and later completely solidifies (Gafur et al, 2003). For the numerical analysis of heat transfer problem, the appropriate set of equations were determined that described the heat transfer behaviour in the cast metal (Hearn, 1992). With the boundary conditions, initial conditions, and thermo-physical properties of the materials being known, it is possible to obtain the temperature and variation of the whole casting system (Ozisik, 1985) and (Liu et al (1993). Finite number at discrete points (Adams and Rogers, (1973), Shampire, (1994) and Bayazitoglu and Ozisik, (1988)) within the cast specimen was employed as the numerical method of solution. This method provides the temperature at a discrete number of points in the cast region. In the numerical method, the cast region is defined and divided into discrete number of points. As temperature difference is imposed in the system, heat flows from the high-temperature region to the low-temperature region as shown in figure 1.

To determine the temperature distribution, energy conservation equations were used for each of the nodal points of the unknown temperature at the interfaces and the cast regions ((Incropera and Dewitt, 1985) and (Janna, 1988)). Temperatures were monitored at distance 2mm into the cast metal, represented by grid point M, and at the steel mould/cast metal interface (see figure 4). By using measured temperatures in both the casting and the steel mould, together with the numerical solutions of the solidification problem, heat transfer coefficients were determined based on Beck (1970) solution of the inverse heat conduction problem. The estimation of the surface heat transfer coefficients or heat flux density utilizing a measured temperature history inside a heat-conducting solid is called the inverse heat conduction problem (Cho and Hong 1996). This problem becomes non-linear, as the thermal properties (thermal conductivity, specific heat) are temperature dependent.

14. Experimental procedure

Chromel-Alumel thermocouples TC2, TC3, TC4, TC5 and TC6 were positioned on the sides of the cylindrical steel container, while TC1 and TC7 were positioned in the cast aluminium metal in the cylindrical and bottom flat surfaces respectively as shown in figure 4 below.

Thermocouples of chromel-alumel type, 3mm in diameter were used to determine the solidifying temperatures of the cast molten metal and heating temperatures of the steel mould at the various positions in the cylindrical steel container of figure 4. The solidifying temperatures at both the cylindrical and flat bottom surfaces of the cast molten aluminium metal were monitored at a position 2mm (from the surface of the steel mould –cast aluminium metal interface) into the cast molten aluminium metal.

At the steel mould wall in the cylindrical surface, thermocouples were positioned at X2 = 4mm, X3 = 8mm, X4 = 12mm, X5 = 16mm and X6 = 20mm measured from the cast aluminium metal / steel mould interface to monitor the heating temperatures at these positions of the steel mould wall as shown in fig. 4. From the temperatures versus time curves obtained for each position in the steel mould, the interface heating temperature versus time curve at the cast aluminium metal / steel mould, for position when X = 0 was obtained by using the polynomial curve fitting method.

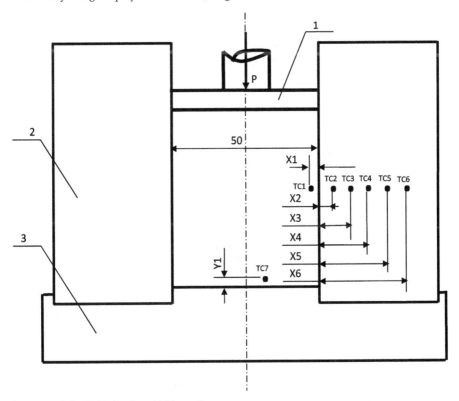

1-upper punch, 2-cylindrical steel mould, 3-lower die
(X1 = 2, X2 = 4, X3 = 8, X4 = 12, X5 = 16, X6 = 20, Y1 = 2 (mm))

Figure 4. Schematic diagram of squeeze casting test rig

This was done by selecting a particular time of heating of steel mould, say t = 10sec. and drawing vertical lines cutting across the heating temperatures versus time curves at various thermocouples' distances within the steel mould. At the point of intersection with each curve, the value of temperature was read against distance X, for the chosen time, t = 10sec. The value of interface steel mould / cast temperature at time say, t = 10sec. was determined at the steel mould / cast metal interface by substituting the value of X = 0 in the polynomial curve fitting equation (70) obtained from the values of temperatures at various distances in the steel mould at a chosen time, t = 10sec.

$$T_{X0} = 0.0031X^4 - 0.168X^3 + 3.263X^2 - 22.812X + 117.8 \tag{70}$$

The temperature obtained by this method corresponds to the interface steel mould / cast metal temperature at a distance X = 0 for the chosen time t = 10sec. If this procedure is repeated for a number of time increments, the temperatures obtained with corresponding times represent the temperature at X=0, for such time increments. The graph of extrapolated temperatures versus time is drawn for position when X = 0 to represent the heating temperatures versus time curve at the steel mould / cast aluminium metal interface is shown in figure 5.

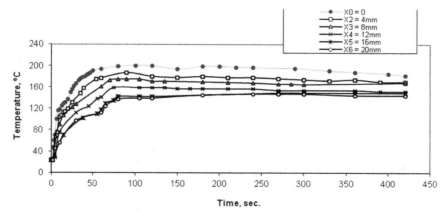

Figure 5. Effect of distance on the heating temperatures of steel mould (extrapolated heating curve at the cast specimen/steel mould interface i.e. X=0)

15. Interface heat transfer coefficients determination

Extrapolated temperature versus time curve of figure 5 for position when X = 0 (i.e. cast aluminium metal / steel mould interface) was used to determine the heat transfer coefficients of solidifying molten aluminium metal. It was used to determine the interface heat transfer coefficients in the cast aluminium metal / steel mould for no pressure and with pressure applications at both the cylindrical and bottom flat surfaces of the steel mould.

The interface heat transfer coefficients between the steel mould and cast aluminium metal at the cylindrical and bottom flat surfaces were determined from the extrapolated experimental heating temperature versus time curve obtained for position X = 0 and aluminium cast solidification temperature versus time curves obtained for the cylindrical and bottom flat surfaces, using equations (66) and (67).

The interface heat transfer coefficients were determined also numerically by the inverse method using the Finite Difference Method (FDM) and the obtained results were compared with the experimentally derived values.

16. Discussions of results

16.1. Temperature-time curves

Figure 6 shows typical temperature versus time curves for solidifying molten aluminium metal and steel mould respectively without the application of pressure on solidifying metal.

This figure shows the comparison of the numerical method usually applied by Cho and Hong (1996) to determine interface steel mould / cast metal temperature versus time curve with the extrapolated experimental method of this present work. The heating curve, as obtained through extrapolations of polynomial curves fitting equations and numerical methods are in close agreement and the deviations from the values obtained numerically varied from between 1.26- 19.31%.

Typical result obtained under pressure is also shown in figure 7, indicating the solidification and heating curves generated for solidifying molten aluminium metal and steel mould which follow the same patterns to the curves in figure 6.

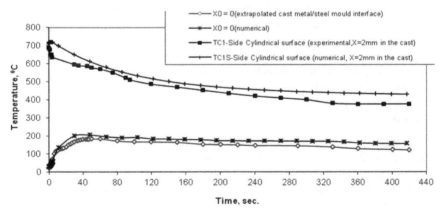

Figure 6. Comparison of experimental measured temperatures with numerical values of aluminium metal without pressure application (P = 0)

With the application of pressure, the peak temperatures recorded are about the same 649°C and 648°C for a pressure of 85.86 MPa at the bottom flat and cylindrical surfaces of the steel mould respectively (see figure 7). The peak temperature (649°C) obtained at the bottom flat surface of the steel mould under applied pressure is found to be higher than that temperature (607°C) without pressure application. This effect may be associated to additional internal heat generated, resulting to higher temperature during pressure application on the solidifying molten aluminium.

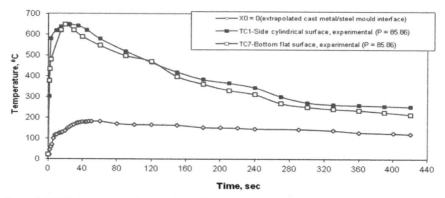

Figure 7. Effect of pressure on the experimental measured temperatures of solidification of aluminium metal (P = 85.86MPa) for side and bottom mould's surface

16.2. Interface heat transfer coefficients with time

From the temperature with time curves of figure 6, the heat transfer coefficients for both cylindrical and bottom flat surfaces were determined for both numerical and calculated values and shown in figure 8. The maximum heat transfer coefficients of 2927.92 W/m²K and 2975.14 W/m²K are obtained at the cylindrical and bottom flat surfaces respectively for no pressure application, which is close to 2900 W/m²K as obtained for pure aluminium by Kim and Lee (1997). The values of heat transfer coefficients decrease rapidly for both the cylindrical and bottom flat surfaces to a level of 866.70 W/m²K and 969.50 W/m²K respectively in 90 seconds. These values further decrease to 361.80 W/m²K and 478.80 W/m²K at these surfaces in another 150 seconds and further decrease then becomes not so noticeable.

From figure 8, the peak values of interface heat transfer coefficients are 2927.92 W/m²K and 2956.73 W/m²K as obtained by experimental and numerical determinations respectively at the cylindrical surface for no pressure application. For times within 40 seconds to 120 seconds the values of the interfacial heat transfer coefficients obtained numerically and experimentally are found to show higher values of about 19.83 % for numerical results to experimental results.

With the application of pressure on the solidifying aluminium metal, the heat transfer coefficients reach maximum values of 3085.34 W/m²K and 3351.08 W/m²K in the cylindrical and bottom flat surfaces respectively (see figure 9). These values also decrease to 847.80

W/m²K and 783.63 W/m²K in 240 seconds in the cylindrical and bottom flat surfaces respectively, while further decrease with time of solidification is no longer pronounced.

16.3. Interface heat transfer coefficients with solidification temperatures

Figures 10 and 12 show the calculated experimental interface heat transfer coefficients for solidifying molten aluminium metal as a function of solidification temperatures of the solidifying molten aluminium metal. Figure 10 shows the variation of heat transfer coefficient with solidification temperatures of aluminium at the cylindrical surface, while figure 12 is the interface heat transfer coefficients with solidifying temperature at the bottom flat surface with and without the application of pressure on the solidifying metal. From the two graphs, the maximum interface heat transfer coefficients obtained without pressure and with pressure application in the bottom flat surface of the steel mould are 2975.14 W/m²K and 3351.08 W/m²K respectively.

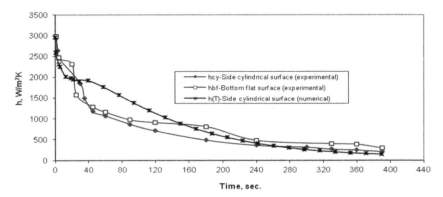

Figure 8. Comparison of numerical values of heat transfer coefficients with calculated experimental values (P = 0)

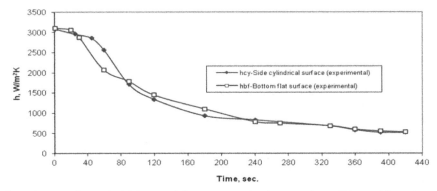

Figure 9. Effect of time of solidification of aluminium metal on heat transfer coefficients with pressure application (P = 85.86MPa) at side and bottom surfaces of steel mould

Figure 11 shows the numerical values of the variation of interface heat transfer coefficients with the solidification temperature of aluminium metal at the cylindrical surface with the application of pressure. The maximum value of heat transfer coefficient of 3397.29 W/m²K at applied pressure of 85.86Mpa as compared to 3351.08 W/m²K obtained through the experimental procedure.

At solidifying temperature above 600°C, a sharp reduction in the interface heat transfer coefficients is noticed at both surfaces as is shown on figures 10, 11 and 12.

For temperatures below 500°C, the interface heat transfer coefficients for both no pressure and pressure applications are close in values. This shows that at temperature below 500°C, the effect of applied pressure is no longer significant on the interface heat transfer coefficient values. The drop in temperature results in solidification of the molten aluminium, which in turn leads to a drop in the heat transfer coefficient values. The effect of applied pressure on the heat transfer coefficients of aluminium becomes more pronounced at solidifying temperatures above 500°C which was also reported by Cho and Hong (1996). Below this temperature, the effect of applied pressure on interface heat transfer coefficient values becomes less pronounced.

Therefore, from figures 10, 11 and 12, it is observed that the effect of applied pressure becomes more significant at temperature close to the liquidus temperature of aluminium as measured along the bottom flat surface of the steel mould (see figure 12). The maximum value of 3397.29 W/m²K is obtained for pressure level of 85.86MPa as compared to 2975.14 W/m²K for no pressure at the bottom flat surface of the steel mould.

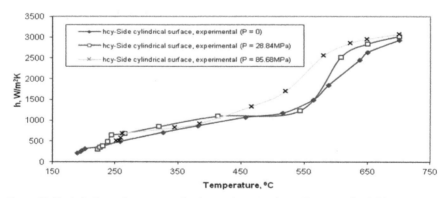

Figure 10. Typical effects of pressure applications on heat transfer coefficients with solidifying temperature at the side cylindrical mould surface by experimental method

16.4. Peak interface heat transfer coefficients with applied pressures

Figure 13 shows the variation of peak values of interface heat transfer coefficients with and without pressure applications. Higher experimental values of heat transfer coefficients are obtained at the bottom flat surface than at the cylindrical surface of the steel mould (see figure 13). This can be associated to greater effect of pressure application experienced at the bottom flat surface than at the side cylindrical surface, thus leading possibly to greater additional internal heat energy generated, and hence obtained higher values of heat transfer coefficients. The results of numerical determination of heat transfer coefficients as in figure 13 shows higher values as compared to the values obtained by experimental method. At applied pressure of 85.86MPa, the obtained heat transfer coefficients are 3397.29 W/m²K and 3351.08 W/m²K by numerical and experimental procedures respectively. From the curves of heat transfer coefficients obtained with temperatures of figures 10, 11 and 12, three distinct portions are noticed. These portions are easily differentiated by the aluminium solidification temperature. These temperatures are below 500⁰C, solidus phase, 500 to 660⁰C, liquidus-solidus phase, and above 660⁰C liquidus phase of solidification of molten aluminium.

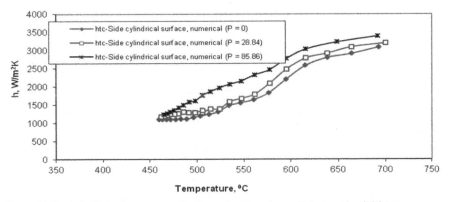

Figure 11. Typical effects of pressure application on heat transfer coefficients with solidifying temperature of the side cylindrical mould surface by numerical method

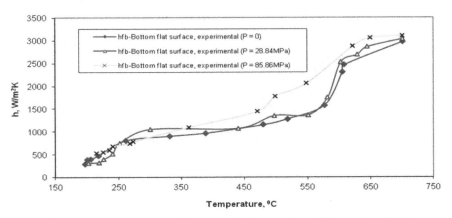

Figure 12. Typical effects of pressure applications on the heat transfer coefficients with solidifying temperature at the bottom flat mould surface

The liquidus-solidus phase, which occurs over a solidification temperature range of 500°C to 660°C, instead of a constant solidification temperature of 660°C can be attributed to the presence of impurities such as silicon, magnesium and manganese in the commercially pure aluminium a fact supported by Higgins (1983).

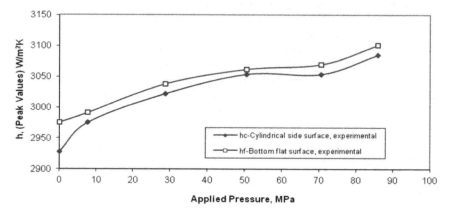

Figure 13. Effect of pressure on the peak values of heat transfer coefficients of aluminium metal at liquidus stage at side and bottom flat moulds' surfaces

The empirical equations for each of the distinct phase changes as a function of applied pressures and solidification temperatures are determined for both the experimental and numerical methods. The empirical equations obtained are for mean average values of heat transfer coefficients based on the experimental method at the cylindrical and flat bottom surfaces. These are equations (71-73):

Temperatures below 500°C (solidus phase),

$$h_{S(exp)} = 3.081T + 1.303P - 232.942 \qquad (71)$$

with coefficient of correlation r = 0.9545.

Temperatures between 500°C and 660°C (liquidus-solidus phase),

$$h_{LS(exp)} = 10.420T + 5.641P - 4176.022 \qquad (72)$$

with coefficient of correlation r = 0.9884.

Temperatures above 660°C (super heat, liquidus phase),

$$h_{L(exp)} = 2.769T + 2.518P + 988.921 \qquad (73)$$

with coefficient of correlation r = 0.7825.

The empirical equations (74-76) obtained through the numerical methods are from the results of the computer simulations of heat transfer coefficients at the cylindrical cast metal /

steel mould interface by the application of various applied pressures. These empirical equations are:

Temperatures below 500°C (solidus phase),

$$h_{S(num)} = 3.849T + 3.643P - 700.427 \qquad (74)$$

with coefficient of correlation r = 0.969

Temperatures between 500°C and 660°C (liquidus-solidus phase),

$$h_{LS(num)} = 9.027T + 3.414P - 3000.625 \qquad (75)$$

with coefficient of correlation r = 0.964

Temperatures above 660°C (super heat, liquidus phase),

$$h_{L(num)} = 2.489T + 2.787P + 1342.19 \qquad (76)$$

with coefficient of correlation r = 0.772

16.5. Die heating effect

Figure 14 is the effect of die pre-heat temperatures on the values of heat transfer coefficients of aluminium metal without the application of pressure. From the figure, the heat transfer coefficients become lower with increase in die pre-heat temperatures. At die temperature of 95°C, the heat transfer coefficient is 3185.34 W/m²K and drop to a value of 2476.73 W/m²K at die temperature of 300°C. For all the die temperatures, there is a fall in the heat transfer coefficient's values as solidification temperature decreases.

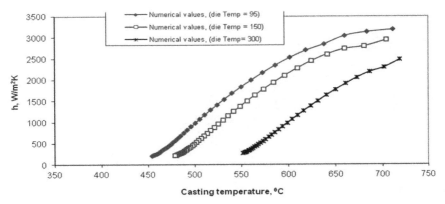

Figure 14. Typical effect of die temperature on heat transfer coefficients without the application of pressure (P=0)

16.6. Comparison of heat transfer coefficient with semi-empirical method

The values of heat transfer coefficients determined using experimental, heat differential equations (numerical), and methods of semi-empirical equations are shown in figure 15 under a pressure application of 85.86MPa. From this graph, the peak values of interface heat transfer coefficient are 3358.19 W/m²K and 3198.79 W/m²K as obtained by heat differential and method of semi-empirical equations respectively for a pressure application of 85.86MPa. The heat transfer coefficients' values for the three methods drop with time and are found to be 1708.03, 1976.81 and 1838.72 W/m²K in 100seconds for experimental, differential and methods of semi-empirical equations respectively.

With die temperature of TM=150°C, the peak heat transfer coefficients of 3088.99 W/m²K, 3249.84 W/m²K and 2982.60 W/m²K are obtained for experimental, heat differential and method of semi-empirical equations as shown on figure 16 following the same pattern as in figure 15.

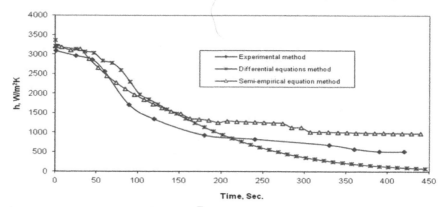

Figure 15. Typical comparison of numerical values of interface heat transfer coefficients with experimental and empirical values with pressure application (P=85.86MPa, TM =300C)

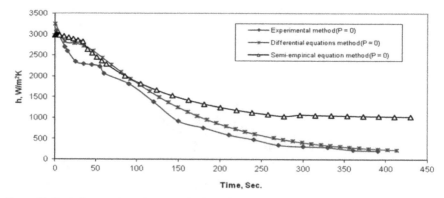

Figure 16. Typical comparison of numerical values of interface heat transfer coefficients with experimental and empirical values with die heating (TM=150°C)

17. Conclusions

The following conclusions can be made from the present investigation:

The graph of temperature against time curves obtained by extrapolating to steel mould / cast metal interface by polynomial curve fitting to heating temperatures graphs with times at various steel mould locations are found to agree in values to the usual numerical methods obtained by previous authors. The interface heat transfer coefficients obtained by the numerical and experimental methods without the application of pressure are found to have values close to that of the numerical methods. The values of the numerical methods were higher by about 19.83%.

Effect of pressure application on the solidifying molten aluminium is more pronounced at casting temperatures above 500^{0}C of the cast aluminium specimen on the values of the interface heat transfer coefficients obtained. Interface heat transfer coefficients are found to decrease with increase in solidification time in both the cylindrical and bottom flat surfaces of the steel mould and thereafter remain fairly constant at temperature below 500^{0}C

Values of experimental peak heat transfer coefficients at the bottom flat surface are found to be higher with pressure application on the solidifying aluminium metal than at the cylindrical surface.

The empirical equations, relating the values of interface heat transfer coefficients with the applied pressures and solidification temperatures at three distinct stages of solidifying molten aluminium are determined and can be applied to determine the heat transfer coefficients.

The values of heat transfer coefficients obtained by heat differential equations incorporating internal heat energy and methods of semi-empirical equations are very close in values. The values as obtained by semi-empirical equations were higher by about 1.7% within the first 5 seconds of solidification.

The semi-empirical equations generated are flexible and could be used to predict the casting temperatures of other metals if the heat transfer coefficient values at the three phase changes are known.

Nomenclature

C_L - Specific heat of liquid molten aluminium metal,
C_S -Specific heat of solidified molten aluminium metal,
C_{st} - Specific heat of steel mould,
d_{sc} -Spatial step size in the solidified molten metal at completion of solidification,
d_L -Spatial step sizes at the Liquid portion,
d_{PS} -Spatial step size at the solidified molten metal phase change,
d_{PL} -Spatial step size at the solidified molten metal phase change,
d_s -Spatial step size in the solidified molten metal,
h_0 -Initial height of cast specimen metal (no pressure application),

h_{PL} -Spatial step size at the liquid molten metal phase change,
h_{PS} -Spatial step size at the solidified molten metal phase change,
H^* -Convective heat transfer of the surrounding atmosphere around the steel mould,
I   -grid point,
G   -grid point at the steel mould-solidified molten metal interface,
M   -grid point at the solidified molten metal-liquid molten metal interface,
N   -grid point at the centre of the solidifying cast specimen metal,
K_L -Thermal conductivity of liquid molten metal,
K_{st} -Thermal conductivity of steel mould material,
K_S -Thermal conductivity of solidified molten metal,
R_o -external radius of steel mould,
L   -internal radius of steel mould,
r   -radial axis/coordinate,
T   -Temperature,
T_∞ -ambient temperature,
T_i -temperature at a specified grid point(i, j),
T_L -temperature of liquid molten metal,
T_{st}, T_M -temperature of steel mould,
T_{MM} -melting temperature of cast (aluminium) metal,
T_P -pouring temperature of liquid molten metal (super heat temperature),
T_S -temperature of solidified molten metal,
X^i_r -thickness of solidified molten metal at a specified grid point-radial direction,
Q   -Thickness of the cylindrical steel mould.

Greek symbols

α_L -thermal diffusivity of the liquid molten metal,
α_{st} -thermal diffusivity of the steel mould material,
α_S -thermal diffusivity of solidified molten metal,
δ -time interval step,
ρ_{st} -density of steel mould material,
ρ_S -density of solidified molten metal,
ρ_L -density of liquid molten metal,
$\sigma_{(T)}$ -Plastic flow stress (temperature dependent).

Author details

Jacob O. Aweda and Michael B. Adeyemi
Department of Mechanical Engineering, University of Ilorin, Ilorin, Nigeria

18. References

Adams, J. Alan & Rogers, David F., (1973), "Computer Aided Heat Transfer Analysis", McGraw-Hill Publishing Company, Tokyo.

Bayazitoglu, Yildiz and Ozisik, M. Necati, (1988), "Elements of Heat Transfer", McGraw-Hill Book Company, New York.

Beck J.V., (1970), "Nonlinear Estimation Applied to the Nonlinear Inverse Heat Conduction Problem", Int. J. Heat Mass Transfer, vol.13, pp703-716.

Bolton, W., (1989), "Engineering Materials Technology", Butterworths-Heinemann Limited, UK.

Browne, David J. and O'Mahoney, D., (2001), "Interface Heat Transfer in Investment casting of Aluminium", Metallurgical and Materials Transactions A, Dec. Vol.32A, pp3055-3063.

Callister, William D. Jr., (1997), "Materials Science and Engineering: An Introduction", 4th Edition, John Wiley and Sons Inc.

Chattopadhyay, Himadri, (2007), "Simulation of transport process in squeeze casting", J. Materials Processing Technology, 186, pp174-178.

Cho I. S. and Hong C. P., (1996), "Evaluation of Heat-Transfer Coefficients at the Casting/Die Interface in Squeeze Casting", Int. J. Cast Metals Res., v.9, pp227-232.

Das A., and Chatterjee S. (1981), "Squeeze Casting of an Aluminium Alloy Containing Small Amount of Silicon Carbide Whiskers", The Metallurgist and Materials Technologist, pp137-142.

Elliott, R., (1988), "Cast Iron Technology", Butterworths, London, U.K.

Gafur, M.A., Nasrul Haque and K. Narayan Prabhu, (2003), "Effects of chill Thickness and Superheat on Casting/Chill Interfacial Heat Transfer During Solidification of Commercially Pure Aluminium", J Materials Processing Technology, 133, pp257-265.

Hearn, E.J. (1992), "Mechanics of Materials", 2nd Edition, Pergamon Press, UK.

Higgins, Raymon A., (1983), "Engineering Metallurgy Part I: Applied Physical Metallurgy", 6th Edition, ELBS with Edward Arnold, UK.

Hosford, William F. and Caddell, Robert M., (1993), "Metal Forming, Mechanics and Metallurgy", 2nd edition, PTR Prentice-Hall Englewood, NJ.

Hu H, and Yu, A., (2002), "Numerical simulation of squeeze cast magnesium alloy AZ91D", Modelling Simul. Mater Sci. Eng., vol.10, pp1-11.

Incropera, Frank P. and Dewitt, David P., (1985), "Fundamental of Heat and Mass Transfer", 3rd Edition, John Wiley and Sons, NY.

Janna, William S., (1988), "Engineering Heat Transfer", SI Edition, Van Nostrand Reinhold (International0, U.K.

Kim, T.G. and Lee, Z.H. (1997), Time-varying heat transfer coefficients between tube-shaped casting and metal mold, Int J, heat mass transfer, 40(15), pp3513-3525.

Kobryn P.A. and Semiatin S. L., (2000), "Determination of Interface Heat-Transfer Coefficients for Permanent Mold casting of Ti-6AL-4v", Metallurgical and materials Transactions, August, Vol.32B, pp685-695.

Liu, A., Voth, T. E. and Bergman, T. L., (1993), "Pure Material Melting and Solidification with Liquid Phase Buoyancy and Surface Tension Forces", Int. J. Heat Mass Transfer, Vol.36 No 2, pp441-442.

Maleki, A., Niroumand, B. and Shafyei, A, (2006), "Effects of squeeze casting parameters on density, macrostructure and hardness of LM13 alloy", Materials Science and Engineering A, 428 pp135-140.

Martorano M. A. and Capocchi J.D.T., (2000), "Heat Transfer Coefficient at the Metal-Mold Interface in the Unidirectional Solidification of Cu-8%Sn alloys", Intl J. Heat Mass Transfer, 43, pp2541-2552.

Ozisik, M. Necati, (1985), "Heat Transfer: – A Basic Approach", McGraw-Hill Publishing, Company U.K.

Potter, D.A. & Easterling, K.E., (1993), "Phase Transformations in Metals", 2nd Edition, Chapman & Hall, London.

Reed-Hill, R E. and Abbaschian, R., (1973), "Physical Metallurgy Principles", 3rd Edition, PWS-KENT Publishing Company, Boston.

Santos, C.A., Quaresma, J.M.V. and Garcia, A., (2001), "Determination of Transient Heat Transfer Coefficients in Chill Mold Castings", Journal of Alloys and Compounds, 139, pp 174-186.

Santos, C.A., Garcia, A., Frick, C.R and Spim J.A., (2004), "Evaluation of heat transfer coefficients along the secondary cooling zones in the continuous casting of steel billets, Inverse problems, Design and Optimization symposium", Rio de Janeiro, pp1-8

Shampire, Lawrence F., (1994), "Numerical Solution of Ordinary Differential Equations", Chapman & Hall, New York.

White, Frank M., (1991), "Heat and Mass Transfer", Addison Wesley Publishing Company, Massachusetts.

Convection Heat Transfer

Natural Convection Heat Transfer from a Rectangular Block Embedded in a Vertical Enclosure

Xiaohui Zhang

Additional information is available at the end of the chapter

1. Introduction

In many circumstances of practical concern, thermal sources are encapsulated into closed cavities containing a fluid, such as in the case of fuel tanks. In other applications it is the heat source itself which needs to be thermally controlled, such as in electronic packaging, passive cooling, space heating, nuclear design, and geophysics; another example is the natural convection around a horizontally-placed or vertically-positioned radiator, which can be used for a centralized heating and cooling system to regulate the air temperature in a cavity. The location of the radiator affects the temperature distribution and heat transfer in a cavity. Normally, a higher position of the radiator is reasonable when the radiator is used as a cooling device, while a lower position for the radiator is reasonable when the radiator is used for a heating one. It is worth studying the temperature distribution and heat transfer in thermal management and design. Whether the radiator is used as a cooling device or a heating device, the heat transfer of a object in a cavity can be simplified and dominated by natural convection heat transfer mechanism in an enclosure with an isolated plate.

Shyy and Rao [1] conducted an investigation of transient natural convection around an enclosed vertical plate. Numerical simulations and experimental data of natural convection air cooling of an array of two-dimensional discrete flush heaters on a vertical wall of a rectangular enclosure were performed by Ho and Chang [2]. Yang and Tao [3] developed a computational method to deal with the internal isolated islands (set the main diagonal element coefficient big values in velocity discrete equations)for natural convection in an enclosure. Experimental work and numerical simulation were studied by Wang [4] regarding natural convection in an inclined cube enclosure with multiple internal isolated plates. Also, numerical analysis on a 3×3 array of discrete heat sources flush-mounted on one vertical wall of a rectangular enclosure filled with various liquids was done by Tou and

Tso [5]. Natural convection from a discrete bottom flush-mounted rectangular heat source on bottom of a horizontal enclosure was studied by Sezai and Mohamad [6]. Deng et al. [7] investigated numerically the steady state natural convection induced by multiple discrete heat sources (DHSs) in horizontal enclosures. Unsteady convection numerical modelling in a vertical channel with a square cylinder was studied by Saha [8], Static bifurcation was found by Liu and Tao [9,10] who performed numerical computations for the heat transfer and fluid flow characteristics of an internal vertical channel composed by a pair of parallel plates situated in an enclosure. Barozzi and Corticelli [11,12] investigated the two-dimensional buoyant flow in a closed cabinet containing two vertical heating plates with a time-accurate finite method. The predictions showed the long-term behavior of numerical solution is time-dependent. The studies mentioned above have not been concerned with the effect of location for heat source or heat sink on the fluid flow and heat transfer.

Following the pioneering numerical works mentioned above, the present study represents a further effort to extend the studies with numerical simulation. The main objective of this study is to analyze the variation effect of the horizontal and vertical location ratios, a/H and b/H, respectively (defined in the following section) for different cold and hot isolated vertical plates in an enclosure, with respect to the flow configuration, temperature distribution, temperature difference distribution (defined in the following section) and heat transfer characteristics of the natural convection. Effect of Rayleigh number on the fluid flow and heat transfer is also presented.

2. Physical model and numerical method

The physical configuration and boundary conditions of problems investigated in this study are shown in Fig. 1. The two horizontal walls are considered to be insulated, and the two vertical walls which have the same temperatures, as well as the isolated vertical plate are maintained at T_1 and T_2, respectively. In this study, the investigations are carried out through the variation of horizontal location ratio a/H and vertical location ratio b/H in the cases of $T_1 > T_2$ (in this case $T_1 = T_h, T_2 = T_c$) and $T_1 < T_2$ (in this case $T_1 = T_c, T_2 = T_h$).

In the present model, the flow is simulated as a two dimensional phenomenon with the following assumptions or simplifications: a) the fluid (air) is Newtonian, incompressible and the flow is laminar, and; b) the temperature difference $T_1 - T_2$ is small, so that the effect of temperature on fluid density is expressed adequately by the Boussinesq approximation. Next, we consider the following dimensionless variables:

$$X = \frac{x}{H}; Y = \frac{y}{H}; U = \frac{u}{U_R}; V = \frac{v}{U_R}; P = \frac{p}{\rho U_R^2}; \Theta = \frac{T - T_c}{T_h - T_c}; \Pr = \frac{v}{a}; Ra = \frac{g\beta H^3 \Delta T}{va}.$$

where the reference velocity is defined as $U_R = (Ra\Pr)^{1/2} a / H$.

The governing equations, that express the conservation of mass, momentum and energy in the fluid domain, become:

$$\frac{\partial U}{\partial X} + \frac{\partial V}{\partial Y} = 0 \tag{1}$$

$$U\frac{\partial U}{\partial X} + V\frac{\partial U}{\partial Y} = -\frac{\partial P}{\partial X} + \frac{Pr}{(RaPr)^{1/2}} \cdot \nabla^2 U \tag{2}$$

$$U\frac{\partial V}{\partial X} + V\frac{\partial V}{\partial Y} = -\frac{\partial P}{\partial Y} + \frac{Pr}{(RaPr)^{1/2}} \cdot \nabla^2 V + \Theta \tag{3}$$

$$U\frac{\partial \Theta}{\partial X} + V\frac{\partial \Theta}{\partial Y} = \frac{K}{(RaPr)^{1/2}} \cdot \nabla^2 \Theta \tag{4}$$

where $\nabla^2 = \partial^2 / \partial X^2 + \partial^2 / \partial Y^2$ and $K = k/k_f$, with k and k_f being the vertical block location conductivity and thermal conductivity of the fluid, respectively.

No-slip condition is imposed on all the walls for the velocities. Thermal boundary conditions are that $\partial \Theta / \partial Y = 0$ for the horizontal insulated walls, and $\Theta = 1$ for the inner heat source and $\Theta = 0$ for the vertical cold walls or $\Theta = 0$ for the inner heat sink and $\Theta = 1$ for the vertical hot walls.

The average Nusselt number is given below:

$$Nu = \frac{H}{2(H+W)} \left(\int_0^{L/H} -\frac{\partial T}{\partial X}\Big|_{X=0} dY + \int_0^{L/H} \frac{\partial T}{\partial X}\Big|_{X=L/H} dY \right) \tag{5}$$

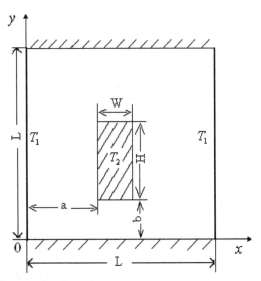

Figure 1. Schematic diagram of configuration

Equations (1) to (5) are solved using a finite volume method (FVM) on a staggered grid system [13]. In the course of discretization, QUICK scheme is adopted to deal with convection and diffusion terms. The equations from the discretization of Eqs. (1) to (4) are solved by the line-by-line procedure, combining the tri-diagonal matrix algorithm (TDMA) and successive over-relaxation iteration (SOR) and the Gauss-Seidel iteration technique with additional block-correction method for fast convergence. The SIMPLE algorithm [13] is used to treat the coupling of the momentum and energy equations. Pressure–correction and velocity-correction schemes are implemented in the model algorithm to arrive at converged solution when both the pressure and velocity satisfy the momentum and continuity equations. The solution is considered to converge when the sum of the normalized residuals for each control equation is of order 10^{-6}.

Special attention is paid to treatment of the isolated solid region. The presence of isolated area is accounted for by a strategy [3, 14] in which a part of solution domain is located in the flow field, therefore, the governing Eqs. (1) to (4) apply to both the fluid and the solid regions. Both the velocity and the dimensionless temperature in it remained zero in iteration process. For details, Ref.[3] may be consulted.

Non-uniform staggered grid system is employed with denser grids clustering near the plate and walls so as to resolve the boundary layer properly. Test runs are performed on a series of non-uniform grids to determine the grids size effects for the Rayleigh numbers 10^4, 10^5 and 10^6 at grid systems 40×40, 60×60 and 80×80, respectively. For each calculation case, a grid independent resolution is obtained. The maximum difference in average Nusselt number between grid (40×40) and grid (60×60) is 5%, the difference in average Nusselt number between grid (60×60) and grid (80×80) is less than 0.2%, so the 60×60 non-uniform grids are used.

The developed computational model is validated against benchmark computational results and is also compared with experimental data. Accuracy of the numerical procedure is validated by the comparison between the predicted results with the benchmark solutions of de Vahl Davis [15] for pure natural convection model in a square cavity with opposite heated and cooled side walls. As shown in Table 1, good agreements are achieved for both the maximum velocity and the average Nusselt number in a broad range of Rayleigh numbers Ra=10^4 to 10^6. Another comparison is also made with respect to the experimental work of Wang [4]. The computationally obtained flow patterns for $Ra = 2 \times 10^5$ are compared with the flow visualization in Fig. 2. It is seen that the model adequately predicts the flow patterns obtained in the visualizations.

Ra	Present		Benchmark	
	U_{max}	Nu	U_{max}	Nu
10^4	0.192	2.231	0.193	2.245
10^5	0.129	4.50	0.132	4.510
10^6	0.078	8.817	0.077	8.806

Table 1. Comparisons between the predicted results and the benchmark resolutions of de Vahl Davis[15]

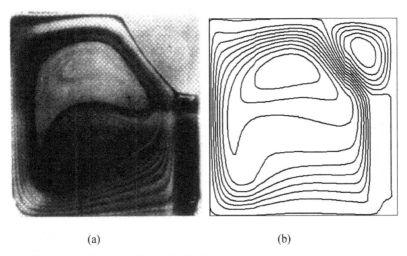

(a)   (b)

Figure 2. Flow field (a) Flow visualization [4], (b) Computation

3. Results and discussion

The configuration dimension is in the proportion of L:H:W = 80 : 15 : 3 .We will examine the fluid flow and heat transfer characteristics under the circumstances of heating or cooling at various locations, followed by the dependence to the Rayleigh number.

3.1. Effect of location on the fluid flow and heat transfer

Numerical simulations have been conducted to elucidate the effect of location ratios variation a/H (0.3≤ a/H ≤2.7, b/H =0.3) and b/H (0.3≤ b/H ≤4.0, a/H =0.3) on the natural convection of a cold isolated vertical plate and hot isolated vertical plate in an enclosure, Rayleigh number is fixed at $Ra = 2.5 \times 10^4$ in both cases.

3.2. Flow and temperature fields

The buoyancy-driven flow and temperature fields in the enclosure for a cold plate and a hot plate with the variation of location ratios a/H and b/H are illustrated by means of contour maps of streamlines and isotherms, respectively, as exemplified in Figs. 3 to 6.

For Fig.3 (a-f), there are two vortices flow where the directions are different for cold plate and hot plate. With the increase of horizontal location ratio a/H near the middle of the enclosure, the two vertices structure around the vertical plate tend to be more symmetric for cold plate and hot plate. The spaces at the two sides of the plate are large enough for the flow within it to develop, so buoyant convection flow is strong on the upper surface of hot plate, indicating the strong effect of the natural convection, while convection is weak in the case of cold plate relatively.

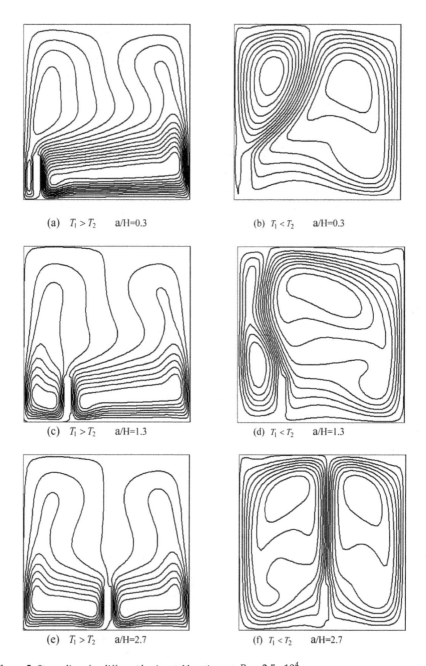

(a) $T_1 > T_2$ a/H=0.3

(b) $T_1 < T_2$ a/H=0.3

(c) $T_1 > T_2$ a/H=1.3

(d) $T_1 < T_2$ a/H=1.3

(e) $T_1 > T_2$ a/H=2.7

(f) $T_1 < T_2$ a/H=2.7

Figure 3. Streamlines for different horizontal locations at $Ra = 2.5 \times 10^4$

Dimensionless temperature distributions, plotted against the variation of the horizontal location ratio a/H , are displayed in Fig.4 (a-f). From this plot, it can be seen that isotherms near walls are almost vertical up to the upper and lower walls. This is due to the conduction effect. The figures reveal also that dimensionless temperature distributions of cold plate are more stratified than those of the hot plate.

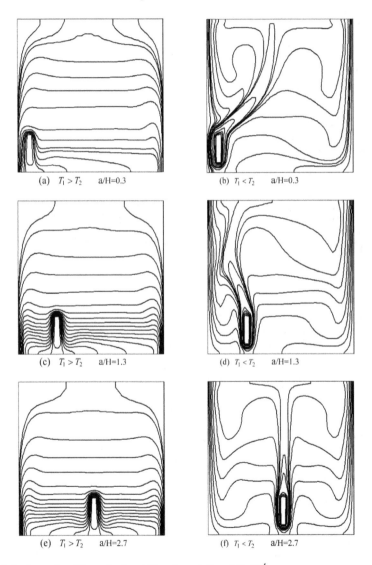

(a) $T_1 > T_2$ a/H=0.3   (b) $T_1 < T_2$ a/H=0.3

(c) $T_1 > T_2$ a/H=1.3   (d) $T_1 < T_2$ a/H=1.3

(e) $T_1 > T_2$ a/H=2.7   (f) $T_1 < T_2$ a/H=2.7

Figure 4. Isotherms for different horizontal locations at $Ra = 2.5 \times 10^4$

For Fig. 3(a and b), Fig. 5 and Fig. 6, we observe significantly different fluid flow and temperature distribution phenomena: the larger of the vertical location ratio b/H, the stronger the flow below the cold plate surface, and the more stratified temperature distribution for the hot plate.

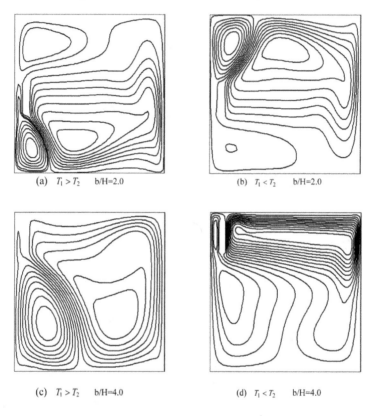

    (a) $T_1 > T_2$ b/H=2.0   (b) $T_1 < T_2$ b/H=2.0

    (c) $T_1 > T_2$ b/H=4.0   (d) $T_1 < T_2$ b/H=4.0

Figure 5. Streamlines for different vertical locations at $Ra = 2.5 \times 10^4$

It is apparent from the comparison of fluid flow configurations and dimensionless temperature distributions of the cold plate and hot plates that the vertical location ratio b/H has a substantial effect on the flow configuration and temperature profile.

3.3. Temperature difference distribution

Uniform temperature difference in an enclosure may have practical importance, particularly in electronics and air conditioning. In order to describe the uniformity of temperature difference in an enclosure we define TD, which means difference of node temperature and average temperature where average temperature is obtained by a weighted area method.

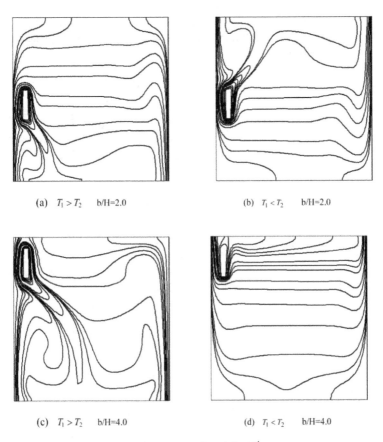

(a) $T_1 > T_2$ b/H=2.0   (b) $T_1 < T_2$ b/H=2.0

(c) $T_1 > T_2$ b/H=4.0   (d) $T_1 < T_2$ b/H=4.0

Figure 6. Isotherms for different vertical locations at $Ra = 2.5 \times 10^4$

$$TD = t_{ij} - \frac{\sum t_{ij} A_{ij}}{\sum A_{ij}}$$ (6)

Two cases are studied at $Ra = 2.5 \times 10^4$, case 1: a/H =0.3 and b/H =0.3, and case2: a/H =0.3 and b/H =2.0. Figure 7 shows the temperature difference distribution, marked as TD and plotted against the height of enclosure, while the width is fixed for heating and cooling conditions.

It can be seen that, in case 1, the temperature difference of heating condition along the height of enclosure is more uniform (the maximum dimensionless temperature difference is 0.11) than that for cooling condition (the maximum dimensionless temperature difference is 0.48). While with the increase of height, the dimensionless temperature differences between heating condition (the maximum dimensionless temperature difference is 0.203) and cooling condition (the maximum dimensionless temperature difference is 0.245) are not great along the height of enclosure in case 2. It implies that the requirement of heating and cooling in case 2 for same component can be met.

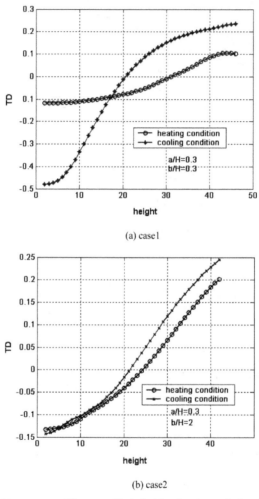

(a) case1

(b) case2

Figure 7. Variation of temperature difference with the height of enclosure for heating and cooling condition

3.4. Heat transfer

Next, attention is focused on the role which the location ratio can play in the heat transfer rate. Figures 8 and 9 present results of the average Nusselt number defined by Equation (5).

For both the cold and the hot plates, b/H is fixed at 0.3. Results show that the same trend occurs with the increase of a/H for both cold plate and hot plate. When the plate is within a wide range of enclosure values, 0.3< a/H <4.8, the effect of its location on heat transfer is small. The variation character of Nu vs. a/H implies that in a wide range of a/H, the spaces at the two sides of the plate are large enough for the flow within it to develop, therefore, the average heat transfer rate is not sensitive to the distance from the side wall. Indeed, the average hot plate Nusselt number is 20% to 39% higher as compared to that of the cold plate. The fact that the average Nusselt number of the hot plate is higher than the corresponding cold plate, can be explained if we consider the streamlines shown in Fig. 3 and temperature profiles in Fig. 4. From these figures we note that the natural convection occurs easier for the hot plate, due to a higher heat transfer than for the cold plate under the same boundary conditions. The steep increase of average Nusselt number near walls for the cold and hot plates in Fig. 8 is referred to as the "chimney effect". The present results confirm the fact that the stronger chimney effect enhances the heat transfer [16, 17]. For Pr ≈1, the boundary layer thickness for the plate scales as [18],

$$\delta = \delta_T \approx y \left(\frac{g\beta\Delta T y^3}{a\nu} \right)^{-1/4} \tag{7}$$

Which indicates that δ increases as $y^{1/4}$.

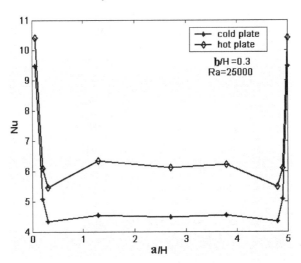

Figure 8. Variation of Nu with horizontal location

The convection is enhanced due to the instability of thermal boundary layers near the left plate and right wall (0< a/H <0.3) or right plate and left wall (4.8< a/H <5). In this spacing, the thermal boundary layers of wall and plate merge with each other, so the boundary layer thickness is thin leading to the increase of Nusselt number. Therefore, the characteristics of the fluid flow and heat transfer in enclosure are very sensitive to the distance between the wall side and the plate.

Different trends occur when the cold plate and the hot plate are at different vertical location which is shown in Fig. 9. Study of Fig. 9 reveals that for cold plate, as the vertical location ratio increases from 0.3 to 3.3 the average Nusselt number increases from 4.33 to 5.68, and slightly decreases from 5.68 to 5.52 when location ratio increases to 4. That means there exists a maximum average Nusselt number when cold plate location $(b/H)_{opt}$ is 3.3. For the

hot plate, average Nusselt number slightly increases from 5.47 to 5.61 when location ratio increase from 0.3 to1.33, and decreases from 5.61 to 4.3 when vertical location ratio increases from 1.33 to 4. There is an appropriate location which corresponds to a maximum heat transfer density for the hot plate also. This can be attributed to two different flow patterns. However, the average Nusselt number variation is not significant for cold and hot plates.

By comparison of Fig. 3(a and b), Fig. 5(c and d), Fig. 6(c and d), and Fig. 9, it can be seen that the symmetry phenomenon appears, *i.e.* the identical problems of natural convection in an enclosure [19].

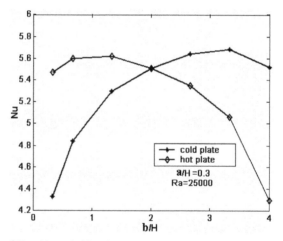

Figure 9. Variation of Nu with vertical location

3.5. Effect of Ra on the fluid flow and heat transfer

The procedure is then repeated over the range $10^2 \leq Ra \leq 10^8$. The effect of the Rayleigh number on the average Nusselt number by setting a/H =0.3 and b/H =0.3 under the circumstances of heating or cooling the block is shown in Fig. 10.

The flow patterns and isotherms are drawn for two typical Rayleigh numbers: $Ra = 1.0 \times 10^2$ and $Ra = 1.0 \times 10^7$ for one geometrical configuration a/H =0.3 and b/H =0.3. These are presented in Figs. 11 and 12.

For the low Rayleigh numbers (Ra= $10^2 - 10^4$), the flow field consists of a single big vortex in the half domain of cavity. With an increase in Rayleigh number, the heat transfer process is dominated successively by conduction mode, combined mode of conduction-convection and convection mechanism. When the Rayleigh number increases to as high as 1.0×10^7, it indicates the convection mode is predominated, these can be seen from Figs. 11 and 12.

At the lower Rayleigh numbers (Ra= $10^2 - 10^4$), Nusselt number of heating condition is a little bit higher than that of cooling condition which can be explained that convection of the hot plate is easier to establish than that of cold plate, conduction mechanism is prevailed and hence heat transfer is mainly dominated by conduction. That is why the Nu values shown in Fig. 10 are almost constant in the low Ra number range.

With the increase in Rayleigh number, Eq. (7) shows that at high Rayleigh numbers, the boundary layers on the enclosure right wall and plate left surface become very thin, leading to a significant increase in Nusselt number. It can be seen from Figs. 10 to 12 that in the convection regime ($Ra > 10^4$), the flow fields difference between the cold plate and the hot plate is appreciable, and the deviations of the Nusselt number between the cold plate and the hot plate are greater. The reason is that heat convection of the hot plate is easier to establish than that of the cold plate.

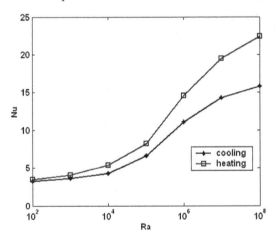

Figure 10. Nu versus Ra for heating and cooling cases

3.6. Special case -non steady state prediction

When the isolate is located in the middle of the enclosure,i.e. $L:H:W:b$=80:15:4:5, After giving definition of dimensionless time $F = \dfrac{a\tau}{H^2}(Ra\,Pr)^{\frac{1}{2}}$, it is easy get the dimensionless non-steady governing equation based on Equation(1)-(4).

The boundary conditions for this system are as same as steady system above.

(a) $Ra = 1.0 \times 10^2$

(b) $Ra = 1.0 \times 10^7$

Figure 11. $T_1 < T_2$ flow patterns (left) and isotherms (right) at different Rayleigh numbers

The zero initial conditions set for velocity and temperature fields.

A perfectly time-periodic solution is predicted shown in Fig. 13 for different Raleigh number($Ra=10^4$, $Ra=5\times10^4$, $Ra=10^5$ and $Ra=10^6$). Fig. 13a reports the time dependent behavior of dimensionless temperature at the monitoring point $(x,y) = (1.3,1.53)$ of the cavity, and the average Nusselt number result is depicted in Fig. 13c. The first noteworthy feature of Fig. 13 is that, except low value of Ra number($Ra=10000,50000$)，after a few time units, a transition to oscillatory flow occurs. the symmetric solution breaks down as instabilities grow, and the time behaviors of quantities relative to geometrically symmetric points begin to differ, The quantities at location $(x,y) = (1.3,1.53)$ exhibit a clearly periodic behavior for Ra=100000(Fig. 13c,d), the period being about 10 time unit.

(a) $Ra = 1.0 \times 10^2$

(b) $Ra = 1.0 \times 10^7$

Figure 12. $T_1 > T_2$ flow patterns (left) and isotherms (right) at different Rayleigh numbers

The isotherms are plotted in Fig. 14 for approximately equal time interval in one periodic circle When analyzing the macroscopic nature of the flow configurations, We observe the isotherms are not centrosymmetric at any point in time. The flow configurations manifest themselves in same pattern—the rolls movie to the left or right with the oscillation.

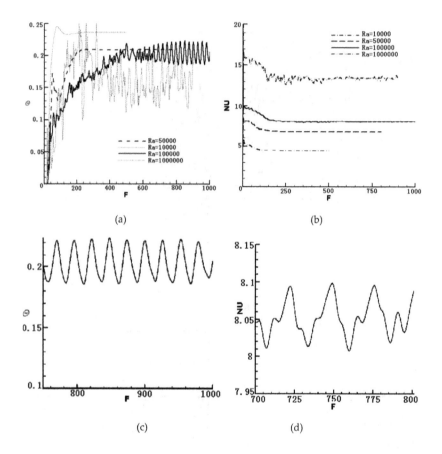

(a)variation of dimensionless temperature with dimensionless time for Ra=10000,50000,100000,1000000 at the monitoring point $(x,y)=(1.3,1.53)$ of the cavity

(b) variation of average Nusselt number with dimensionless time for Ra=10000,50000,100000,1000000 at the monitoring point $(x,y)=(1.3,1.53)$ of the cavity

(c) Partial enlargement of $T-F$ at the monitoring point $(x,y)=(1.3,1.53)$ in the cavity for $Ra=10^5$

(d)Partial enlargement of $Nu-F$ at the monitoring point $(x,y)=(1.3,1.53)$ in the cavity for $Ra=10^5$

Figure 13. Time history of the dimensionless temperature and Nusselt number at the monitoring point $(x,y)=(1.3,1.53)$ of the cavity

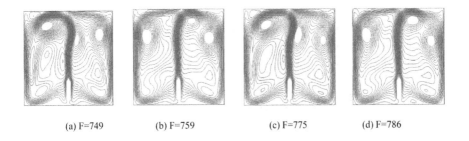

   (a) F=749   (b) F=759   (c) F=775   (d) F=786

Figure 14. Series of instantaneous streamlines for Ra $=10^5$

4. Concluding remarks

A numerical study has been presented to unveil primarily the effect of location ratio variation of cold isolated vertical plate and hot isolated vertical plate on the natural convection in an enclosure. We conclude as follows from the numerical results.

a. The flow configurations and dimensionless temperature profiles of cold plate and hot plate are different;

b. The stratification is found to be strong for cold plate, while for the hot plate is relatively weak;

c. The temperature difference along the height of the enclosure in heating conditions is more uniform than that obtained in cooling conditions;

d. The increase of vertical location near the middle of enclosure leads to small temperature differences in heating and cooling conditions;

e. The trend of the average Nu number variation is the same for the cold and hot plates, when the plate is fixed at different horizontal locations. However, the average Nu number of the hot plate is about 20% to 39% larger than that of cold plate at the same Rayleigh number of 25000. For narrow distances between the inner plate and the bounding wall, the inner plate Nusselt number is enhanced, aside from this, the plate average Nusselt number is insensitive to the plate position;

f. The average Nu numbers tend to decrease with the increase of vertical location ratio (1.33< b/H <4) for hot plate, but for the cold plate the average Nu number tends to augment with the increase of vertical location ratio (0.3< b/H <3.3);

g. An optimum vertical location ratio exists at which the heat transfer is maximum for both cold and hot plates at a specific Rayleigh number;

h. Non-steady modeling simulations reveals that solutions are unique for values of Rayleigh number 10^4 and 5×10^4 where the flow and heat transfer is steady state. While unsteady state flow and heat transfer is appeared as a function of $Ra=10^5$ and $Ra=10^6$ for the rectangular block located in the middle of the cavity.

Nomenclature

a	horizontal location
A	node (i, j) based control volume area
b	vertical location
F	non-dimensional time
g	gravitational acceleration
H	height of vertical plate
k	vertical block location thermal conductivity
k_f	fluid thermal conductivity
K	relative thermal conductivity
L	characteristic length of the enclosure
Nu	Nusselt number
p	pressure
P	non-dimensional pressure
Pr	Prandtl number
Ra	Rayleigh number
t	node (i, j) temperature
T	temperature
TD	temperature difference
u, v	velocity components
U, V	dimensionless velocity
W	width of vertical plate
x, y	Cartesian coordinates
X, Y	dimensionless coordinates

Greek symbols

a	thermal diffusivity
β	coefficient of thermal expansion
δ	velocity boundary layer thickness
δ_T	thermal boundary layer thickness
ρ	density of the fluid
v	kinematic viscosity of the fluid
Θ	dimensionless temperature

Subscipts

c	cold
h	hot
opt	optimization
R	reference

Author details

Xiaohui Zhang

School of Physical Science and Technology, School of Energy, Soochow University, Suzhou, China

Acknowledgement

This work is supported by National Natural Science Foundation of China（No.51176132）

5. References

[1] Shyy, W., and Rao, M.M., Simulation of Transient Natural Convection around an Enclosed Vertical Plate, *ASME J. Heat Transfer*, vol. 115, pp. 946–953, 1993.

[2] Ho, C.J., and Chang, J.Y., A Study of Natural Convection Heat Transfer in a Vertical Rectangular with Two-dimensional Discrete Heating: Effect of Aspect Ratio, *Int. J. Heat Mass Transfer*, vol. 37, pp. 917–926, 1994.

[3] Yang, M., and Tao, W.Q., Numerical Study of Natural Convection Heat Transfer in a Cylindrical Envelope with Internal Concentric Slotted Hollow Cylinder, *Numerical Heat Transfer*, Part A, vol. 22, pp. 281–305, 1992.

[4] Wang, Q.W., Natural Convection in an Inclined Cube Enclosure with Multiple Internal Isolated Plates, Ph.D. Dissertation., Xi'an Jiaotong University,China,1996.

[5] Tou, S.K.W., and Tso, C.P., 3-D Numerical Analysis of Natural Convective Liquid Cooling of 3x3 Heater Array in Rectangular Enclosure, *Int. J. Heat Mass Transfer*,vol.42, pp.3231-3244,1999.

[6] Sezai, I., and Mohamad, A.A., Natural Convection from a Discrete Heat Source on the Bottom of a Horizontal Enclosure, *Int. J. Heat Mass Transfer*, vol. 42, pp. 2257-2266, 2000.

[7] Deng, Q.H., Tang, G.F., and Li, Y.G., Interaction between Discrete Heat Source in Horizontal Natural Convection Enclosures, *Int. J. Heat Mass Transfer*, vol. 45, pp. 5117-5132, 2002.

[8] Saha, A.K., Unsteady free convection in a vertical channel with a built-in heated square cylinder, *Numerical Heat Transfer Part A*, vol. 38, no. 8, pp. 795-818, 2000.

[9] Liu, J.P., and Tao, W.Q., Numerical analysis of natural convection around a vertical channel in a rectangular enclosure. *Heat and Mass Transfer*; vol. 31, no. 5, pp. 313-321,1996.

[10] Liu, J.P., and Tao, W.Q., Bifurcation to oscillatory flow of the natural convection around a vertical channel in rectangular enclosure, *International Journal of Numerical Methods for Heat and Fluid Flow*, vol. 9, no. 2, pp. 170-85,1999.

[11] Barozzi, G.S., and Corticelli, M.A., Natural convection in cavities containing internal heat sources, *Heat and Mass Transfer*, vol. 36, pp. 473-80, 2000.

[12] Barozzi, G.S., and Corticelli, M.A., Nobile E. Numerical simulation of time-dependent buoyant flows in an enclosed vertical channel, *Heat Mass Transfer*, vol. 35, pp. 89-99, 1999.

[13] Patankar, S.V., Numerical Heat Transfer and Fluid Flow, McGraw-Hill, New York, pp. 146-152, 1980.

[14] Tao, W.Q., Numerical Heat Transfer (2ed edition) Xi'an Jiaotong University Press, pp. 244-245, 2001. (in Chinese)

[15] de Vahl Davis G., Natural Convection of Air in a Square Cavity: a Benchmark Numerical Solution, *Int . J. Meth. Fluids*, vol. 3, pp. 249-264, 1983.

[16] Kazansky, S., Dubovsky, V., Ziskind, G., and Letan, R., Chimney-Enhanced Natural Convection from a Vertical Plate: Experiments and Numerical Simulations, *Int. J. Heat Mass Transfer* ,vol. 46, pp. 497-512, 2003.

[17] Auletta, A., and Manca, O., Heat and Fluid Flow Resulting from the Chimney Effect in Symmetrically Heated Vertical Channel with Adiabatic Extensions, *Int. J. of Thermal Sciences*, vol. 41,pp. 1101-1111, 2002.

[18] Bejan, A., Convection Heat Transfer, Wiley, New York, 1984.

[19] Yang , M., Tao, W.Q. and Wang, Q.W., On the Identical Problems of Natural Convection in Enclosures and Applications of the Identity Character, *Int. J. of Thermal Sciences*, vol. 2, pp. 116–25, 1993 .

Chapter 5

Boundary-Layer Flow in a Porous Medium of a Nanofluid Past a Vertical Cone

F.M. Hady, F.S. Ibrahim, S.M. Abdel-Gaied and M.R. Eid

Additional information is available at the end of the chapter

1. Introduction

The natural convection flow over a surface embedded in saturated porous media is encountered in many engineering problems such as the design of pebble-bed nuclear reactors, ceramic processing, crude oil drilling, geothermal energy conversion, use of fibrous material in the thermal insulation of buildings, catalytic reactors and compact heat exchangers, heat transfer from storage of agricultural products which generate heat as a result of metabolism, petroleum reservoirs, storage of nuclear wastes, etc.

The derivation of the empirical equations which govern the flow and heat transfer in a porous medium has been discussed in [1-5]. The natural convection on vertical surfaces in porous media has been studied used Darcy's law by a number of authors [6–20]. Boundary layer analysis of natural convection over a cone has been investigated by Yih [21-24]. Murthy and Singh [25] obtained the similarity solution for non-Darcy mixed convection about an isothermal vertical cone with fixed apex half angle, pointing downwards in a fluid saturated porous medium with uniform free stream velocity, but a semi-similar solution of an unsteady mixed convection flow over a rotating cone in a rotating viscous fluid has been obtained Roy and Anilkumar [26]. The laminar steady nonsimilar natural convection flow of gases over an isothermal vertical cone has been investigated by Takhar et al. [27]. The development of unsteady mixed convection flow of an incompressible laminar viscous fluid over a vertical cone has been investigated by Singh and Roy [28] when the fluid in the external stream is set into motion impulsively, and at the same time the surface temperature is suddenly changed from its ambient temperature. An analysis has been carried out by Kumari and Nath [29] to study the non-Darcy natural convention flow of Newtonian fluids on a vertical cone embedded in a saturated porous medium with power-law variation of the wall temperature/concentration or heat/mass flux and suction/injection. Cheng [30-34] focused on the problem of natural convection from a vertical cone in a porous medium with mixed thermal boundary conditions, Soret and Dufour effects and with variable viscosity.

The conventional heat transfer fluids including oil, water and ethylene glycol etc. are poor heat transfer fluids, since the thermal conductivity of these fluids play an important role on the heat transfer coefficient between the heat transfer medium and the heat transfer surface. An innovative technique for improving heat transfer by using ultra fine solid particles in the fluids has been used extensively during the last several years. Choi [35] introduced the term nanofluid refers to these kinds of fluids by suspending nanoparticles in the base fluid. Khanafer et al. [36] investigated the heat transfer enhancement in a two-dimensional enclosure utilizing nanofluids. The convective boundary-layer flow over vertical plate, stretching sheet and moving surface studied by numerous studies and in the review papers Buongiorno [37], Daungthongsuk and Wongwises [38], Oztop [39], Nield and Kuznetsov [40,41], Ahmad and Pop [42], Khan and Pop [43], Kuznetsov and Nield [44,45] and Bachok et al. [46].

From literature survey the base aim of this work is to study the free convection boundary-layer flow past a vertical cone embedded in a porous medium filled with a nanofluid, the basic fluid being a non-Newtonian fluid by using similarity transformations. The reduced coupled ordinary differential equations are solved numerically. The effects of the parameters governing the problem are studied and discussed.

2. Mathematical formulation of the problem

Consider the problem of natural convection about a downward-pointing vertical cone of half angle γ embedded in a porous medium saturated with a non-Newtonian power-law nanofluid. The origin of the coordinate system is placed at the vertex of the full cone, with x being the coordinate along the surface of the cone measured from the origin and y being the coordinate perpendicular to the conical surface Fig (1). The temperature of the porous medium on the surface of the cone is kept at constant temperature T_w, and the ambient porous medium temperature is held at constant temperature T_∞.

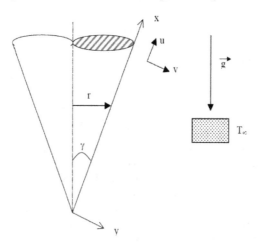

Figure 1. A schematic diagram of the physical model.

The nanofluid properties are assumed to be constant except for density variations in the buoyancy force term. The thermo physical properties of the nanofluid are given in Table 1 (see Oztop and Abu-Nada [39]). Assuming that the thermal boundary layer is sufficiently thin compared with the local radius, the equations governing the problem of Darcy flow through a homogeneous porous medium saturated with power-law nanofluid near the vertical cone can be written in two-dimensional Cartesian coordinates (x, y) as:

$$\frac{\partial(r^m u)}{\partial x} + \frac{\partial(r^m v)}{\partial y} = 0, \tag{1}$$

$$\frac{\partial u^n}{\partial y} = \frac{(\rho\beta)_{nf} Kg \cos\gamma}{\mu_{nf}} \frac{\partial T}{\partial y}, \tag{2}$$

$$u\frac{\partial T}{\partial x} + v\frac{\partial T}{\partial y} = \alpha_{nf} \frac{\partial^2 T}{\partial y^2}. \tag{3}$$

Where u and v are the volume-averaged velocity components in the x and y directions, respectively, T is the volume-averaged temperature. n is the power-law viscosity index of the power-law nanofluid and g is the gravitational acceleration. $m = \gamma = 0$ corresponds to flow over a vertical flat plate and $m = 1$ corresponds to flow over a vertical cone. n is the viscosity index. For the case of $n = 1$, the base fluid is Newtonian. We note that $n < 1$ and $n > 1$ represent pseudo-plastic fluid and dilatant fluid, respectively. Property ρ_{nf} and μ_{nf} are the density and effective viscosity of the nanofluid, and K is the modified permeability of the porous medium. Furthermore, α_{nf} and β_{nf} are the equivalent thermal diffusivity and the thermal expansion coefficient of the saturated porous medium, which are defined as (see Khanafer et al. [36]):

$$\rho_{nf} = (1-\phi)\rho_f + \phi\rho_s, \quad \mu_{nf} = \frac{\mu_f}{(1-\phi)^{2.5}}, \quad \alpha_{nf} = \frac{k_{nf}}{(\rho C_p)_{nf}},$$

$$(\rho C_p)_{nf} = (1-\phi)(\rho C_p)_f + \phi(\rho C_p)_s, \quad \frac{k_{nf}}{k_f} = \frac{(k_s + 2k_f) - 2\phi(k_f - k_s)}{(k_s + 2k_f) + 2\phi(k_f - k_s)}. \tag{4}$$

Here ϕ is the solid volume fraction.

The associated boundary conditions of Eqs. (1)-(3) can be written as:

$$v = 0; \ T = T_w \quad \text{at } y = 0;$$
$$u = 0; \ T \to T_\infty \quad \text{as } y \to \infty, \tag{5}$$

where μ_f is the viscosity of the basic fluid, ρ_f and ρ_s are the densities of the pure fluid and nanoparticle, respectively, $(\rho C_p)_f$ and $(\rho C_p)_s$ are the specific heat parameters of the base fluid and nanoparticle, respectively, k_f and k_s are the thermal conductivities of the base fluid and nanoparticle, respectively. The local radius to a point in the boundary layer r can be represented by the local radius of the vertical cone $r = x \sin \gamma$.

	$\rho(kg/m^3)$	$C_p(J/kgK)$	$k(W/mK)$	$\beta \times 10^5 (K^{-1})$
Pure water	997.1	4179	0.613	21
Copper (Cu)	8933	385	401	1.67
Silver (Ag)	10500	235	429	1.89
Alumina (Al2O3)	3970	765	40	0.85
Titanium Oxide (TiO2)	4250	686.2	8.9538	0.9

Table 1. Thermo-physical properties of water and nanoparticles [39].

By introducing the following non-dimensional variables:

$$\eta = \frac{y}{x} Ra_x^{1/2}, \qquad f(\eta) = \frac{\psi(x,y)}{\alpha_f r^m Ra_x^{1/2}},$$
$$\theta(\eta) = \frac{T - T_\infty}{T_w - T_\infty}. \tag{6}$$

The continuity equation is automatically satisfied by defining a stream function $\psi(x,y)$ such that:

$$r^m u = \frac{\partial \psi}{\partial y} \text{ and } r^m v = -\frac{\partial \psi}{\partial x}. \tag{7}$$

where;

$$Ra_x = \left(\frac{x}{\alpha_f} \right) \left[\frac{Kg(\rho\beta)_f \cos\gamma (T_w - T_\infty)}{\mu_f} \right]^{1/n}. \tag{8}$$

Integration the momentum Eq. (2) we have:

$$\frac{\mu_{nf}}{\mu_f} u^n = \frac{(\rho\beta)_{nf} Kg \cos\gamma}{\mu_f} (T - T_\infty). \tag{9}$$

Substituting variables (6) into Eqs. (1)–(5) with Eq. (9), we obtain the following system of ordinary differential equations:

$$\frac{1}{(1-\phi)^{2.5}} (f')^n = \left[1 - \phi + \phi \frac{(\rho\beta)_s}{(\rho\beta)_f} \right] \theta, \tag{10}$$

$$\frac{k_{nf}/k_f}{\left[1-\phi+\phi\dfrac{(\rho C_p)_s}{(\rho C_p)_f}\right]}\theta'' + \left(m+\frac{1}{2}\right)f\theta' = 0, \qquad (11)$$

along with the boundary conditions:

$$f(0) = 0, \ \theta(0) = 1, \\ f'(\infty) = 0, \ \theta(\infty) = 1. \qquad (12)$$

where primes denote differentiation with respect to η , the quantity of practical interest, in this chapter is the Nusselt number Nu_x which is defined in the form:

$$Nu_x = \frac{hx}{k_m} = \frac{-\dfrac{\partial T}{\partial y}\Big|_{y=0} x}{T_w - T_\infty} = -Ra_x^{1/2}\theta'(0). \qquad (13)$$

where h denotes the local heat transfer coefficient.

3. Results and discussion

In this study we have presented similarity reductions for the effect of a nanoparticle volume fraction on the free convection flow of nanofluids over a vertical cone via similarity transformations. The numerical solutions of the resulted similarity reductions are obtained for the original variables which are shown in Eqs. (10) and (11) along with the boundary conditions (12) by using the implicit finite-difference method. The physical quantity of interest here is the Nusselt number Nu_x and it is obtained and shown in Eqs. (13) and (14). The distributions of the velocity $f'(\eta)$, the temperature $\theta(\eta)$ from Eqs.(10) and (11) and the Nusselt number in the case of Cu-water and Ag-water are shown in Figs. 2–8. The computations are carried for various values of the nanoparticles volume fraction for different types of nanoparticles, when the base fluid is water. Nanoparticles volume fraction ϕ is varied from 0 to 0.3. The nanoparticles used in the study are from Copper (Cu), Silver (Ag), Alumina (Al2O3) and Titanium oxide (TiO2).

In order to verify the accuracy of the present method, we have compared our results with those of Yih [22] for the rate of heat transfer $\theta'(0)$ in the absence of the nanoparticles $(\phi = 0)$. The comparisons in all the above cases are found to be in excellent agreement, as shown in Table 2. It is clear that as a geometry shape parameter m increases, the local Nusselt number increases. While Table 3 depict the heat transfer rate $\theta'(0)$ for various values of nanoparticles volume fraction ϕ for different types of nanoparticles when the base fluid is water. Figs. 2 and 3 show the effects of the nanoparticle volume fraction φ on the velocity distribution in the case of Cu-water when $\phi = 0, 0.05, 0.1, 0.15, 0.2, 0.3$. It is noted that the velocity along the cone increases with the nanoparticle volume fraction in both of the two cases (i.e. Cu-water and Ag-water), moreover the velocity distribution in the case of Ag-water is larger than that for Cu-water. We can show that the change of the velocity distribution when we use different types of nanoparticles from Fig. 4, which depict the Ag-nanoparticles are the highest when the base fluid is water and when $\phi = 0.1$. Thus the presence of the nanoparticles volume fraction increases the momentum boundary layer thickness.

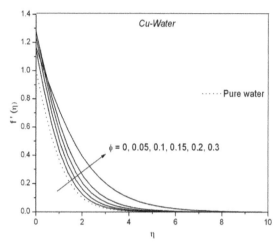

Figure 2. Effects of the nanoparticle volume fraction ϕ on velocity distribution $f'(\eta)$ in the case of Cu-Water.

n	Vertical plate		Vertical cone	
	Yih [22]	Present method	Yih [22]	Present method
0.5	0.3766	0.3768	0.6522	0.6524
0.8	0.4237	0.4238	0.7339	0.7340
1.0	0.4437	0.4437	0.7686	0.7686
1.5	0.4753	0.4752	0.8233	0.8233
2.0	0.4938	0.4938	0.8552	0.8552

Table 2. Comparison of results for the reduced Nusselt number $-\theta'(0)$ for vertical plate $(\lambda = 0)$ and vertical cone $(\lambda = 1)$ when $\phi = 0$.

ϕ	Cu	Ag	Al_2O_3	TiO_2
0.05	0.7423	0.7704	0.6604	0.6725
0.1	0.6931	0.7330	0.5642	0.5852
0.15	0.6301	0.6732	0.4780	0.5057
0.2	0.5591	0.6002	0.4006	0.4331
0.3	0.4052	0.4357	0.2673	0.3062

Table 3. Values of $-\theta'(0)$ for various values of ϕ when $n = 1$.

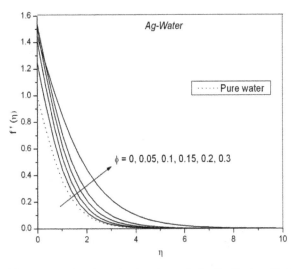

Figure 3. Effects of the nanoparticle volume fraction ϕ on velocity distribution $f'(\eta)$ in the case of Ag-Water.

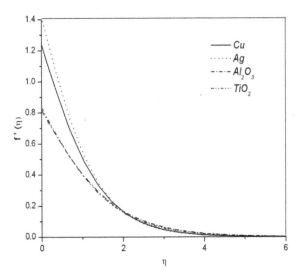

Figure 4. Velocity profiles $f'(\eta)$ for different types of nanofluids when $\phi = 0.1$.

Figs. 5 and 6 are presented to show the effect of the volume fraction of nanoparticles Cu and Ag respectively, on temperature distribution. These figures illustrate the streamline for different values of ϕ, when the volume fraction of the nanoparticles increases from 0 to 0.3, the thermal boundary layer is increased. This agrees with the physical behavior, when the volume of copper and silver nanoparticles increases the thermal conductivity increases, and then the thermal boundary layer thickness increases. Moreover Fig. 7 displays the behavior of the different types of nanoparticles on temperature distribution when $\phi = 0.1$. The figure showed that by using different types of nanofluid as the values of the temperature change and the Ag-nanoparticles are the lower distribution.

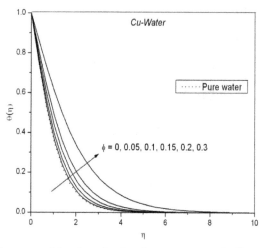

Figure 5. Effects of the nanoparticle volume fraction ϕ on temperature distribution $\theta(\eta)$ in the case of Cu-Water.

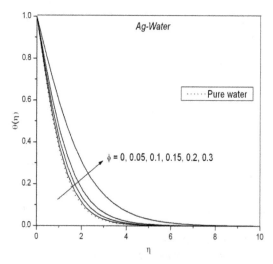

Figure 6. Effects of the nanoparticle volume fraction ϕ on temperature distribution $\theta(\eta)$ in the case of Ag-Water.

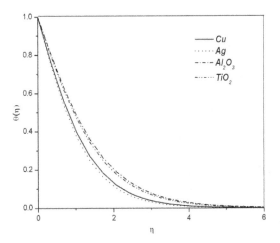

Figure 7. Temperature profiles $\theta(\eta)$ for different types of nanofluids when $\phi = 0.1$.

Fig. 8 shows the variation of the reduced Nusselt number with the nanoparticles volume fraction ϕ for the selected types of the nanoparticles. It is clear that the heat transfer rates decrease with the increase in the nanoparticles volume fraction ϕ. The change in the reduced Nusselt number is found to be lower for higher values of the parameter ϕ. It is observed that the reduced Nusselt number is higher in the case of Ag-nanoparticles and next Cu-nanoparticles, TiO₂-nanoparticles and Al₂O₃-nanoparticles. Also, the Fig. 8 and Table 3 show that the values of $\theta'(0)$ change with nanofluid changes, namely we can say that the shear stress and heat transfer rate change by taking different types of nanofluid. Furthermore this depicts that the nanofluids will be very important materials in the heating and cooling processes.

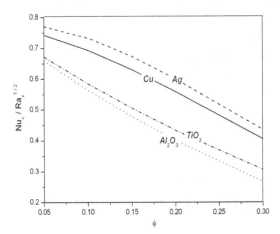

Figure 8. Effects of the nanoparticle volume fraction ϕ on dimensionless heat transfer rates.

4. Conclusions

The problem of the steady free convection boundary layer flow past a vertical cone embedded in porous medium filled with a non-Newtonian nanofluid has been studied and the special case when the base fluid is water has been considered. The effects of the solid volume fraction ϕ on the flow and heat transfer characteristics are determined for four types of nanofluids: Copper (Cu), Silver (Ag), Alumina (Al_2O_3) and Titanium oxide (TiO_2). It has been shown, as expected, that increasing of the values of the nanoparticles volume fraction lead to an increase of the velocity and the temperature profiles and to an decrease of the Nusselt number for the values of the parameter ϕ. It has been found that the Ag-nanoparticles proved to have the highest cooling performance and Alumina-nanoparticles enhanced to have highest heating performance for this problem.

Nomenclature

C_p	specific heat at constant temperature
f	dimensionless stream function
g	acceleration due to gravity
h	local heat transfer coefficient
K	permeability coefficient of the porous medium
k	thermal conductivity
m	geometry shape parameter
Nu_x	reduced Nusselt number
n	viscosity index, $n \geq 0$
Ra_x	modified Rayleigh number
r	local radius of the cone
T	temperature
T_w	temperature at the surface of the cone
T_∞	ambient temperature attained as $y \to \infty$
u, v	Darcian velocity components in x - and y -directions
x, y	Cartesian coordinates

Greek symbols

α	thermal diffusivity
β	volumetric expansion coefficient
γ	half angle of the cone
η	similarity variable
θ	dimensionless temperature
μ	effective viscosity

ρ_f	density of the fluid
ρ_s	nanoparticles mass density
$\left(\rho C_p\right)_{nf}$	heat capacitance of the nanofluid
$\left(\rho C_p\right)_f$	heat capacity of the fluid
$\left(\rho C_p\right)_s$	effective heat capacity of the nanoparticles material
ϕ	nanoparticles volume fraction
ψ	stream function

Subscripts

f	fluid fraction
nf	nanofluid fraction
s	solid fraction
w	condition at the wall
∞	stream function condition at the infinity

Author details

F.M. Hady
Department of Mathematics, Faculty of Science, Assiut University, Assiut, Egypt

S.M. Abdel-Gaied and M.R. Eid*
*Department of Science and Mathematics, Faculty of Education, Assiut University,
The New Valley, Egypt*

F.S. Ibrahim
*Department of Mathematics, University College in Jamoum, Umm Al-Qura University,
Makkah, Saudi Arabia*

5. References

[1] Cheng P (1978) Heat transfer in geothermal systems. *Adv. Heat Transfer* 14: 1–105.
[2] Nield DA, Bejan A (1999) Convection in Porous Media. second ed., *Springer: New York.*
[3] Vafai K (2000) Handbook of Porous Media. *Marcel Dekker: New York.*
[4] Pop I, Ingham DB (2001) Convective Heat Transfer: Mathematical and Computational Modelling of Viscous Fluids and Porous Media. *Pergamon Press: Oxford* .
[5] Ingham DB, Pop I (2002) Transport Phenomena in Porous Media. *Pergamon Press: Oxford*
[6] Cheng P, Minkowycz WJ (1977) Free convection about a vertical flat plate embedded in a porous medium with application to heat transfer from a dike. *J. Geophys. Res.* 82(14): 2040–2044.

* Corresponding Author

[7] Gorla RSR, Tornabene R (1988) Free convection from a vertical plate with nonuniform surface heat flux and embedded in a porous medium. *Transp. Porous Media* 3 : 95–106.

[8] Bakier AY, Mansour MA, Gorla RSR, Ebiana AB (1997) Nonsimilar solutions for free convection from a vertical plate in porous media. *Heat Mass Transfer* 33: 145–148.

[9] Gorla RSR, Mansour MA, Abdel-Gaied SM (1999) Natural convection from a vertical plate in a porous medium using Brinkman's model. *Transp. Porous Media* 36 : 357–371.

[10] Mulolani I, Rahman M (2000) Similarity analysis for natural convection from a vertical plate with distributed wall concentration. *Int. J. Math. math. Sci.* 23(5) : 319 – 334.

[11] Jumah RY, Mujumdar AS (2000) Free convection heat and mass transfer of non-Newtonian power law fluids with yield stress from a vertical plate in saturated porous media. *Int. Comm. Heat Mass Transfer* 27 : 485–494.

[12] Groşan T, Pop I (2001) Free convection over a vertical flat plate with a variable wall temperature and internal heat generation in a porous medium saturated with a non-Newtonian fluid. *Technische Mechanik* 21(4) : 313–318.

[13] Kumaran V, Pop I (2006) Steady free convection boundary layer over a vertical flat plate embedded in a porous medium filled with water at 4°C. *Int. J. Heat Mass Transfer* 49 : 3240–3252.

[14] Magyari E, Pop I, Keller B (2006) Unsteady free convection along an infinite vertical flat plate embedded in a stably stratified fluid-saturated porous medium. *Transp. Porous Media* 62 : 233–249.

[15] Cheng C-Y (2006) Natural convection heat and mass transfer of non-Newtonian power law fluids with yield stress in porous media from a vertical plate with variable wall heat and mass fluxes. *Int. Comm. Heat Mass Transfer* 33 : 1156–1164.

[16] Chamkha AJ, Al-Mudhaf AF, Pop I (2006) Effect of heat generation or absorption on thermophoretic free convection boundary layer from a vertical flat plate embedded in a porous medium. *Int. Comm. Heat Mass Transfer* 33 : 1096–1102.

[17] Magyari E, Pop I, Postelnicu A (2007) Effect of the source term on steady free convection boundary layer flows over an vertical plate in a porous medium. Part I. *Transp. Porous Media* 67 : 49–67.

[18] Nield DA, Kuznetsov AV (2008) Natural convection about a vertical plate embedded in a bidisperse porous medium. *Int. J. Heat Mass Transfer* 51 : 1658–1664.

[19] Mahdy A, Hady FM (2009) Effect of thermophoretic particle deposition in non-Newtonian free convection flow over a vertical plate with magnetic field effect. *J. Non-Newtonian Fluid Mech.* 161: 37–41.

[20] Ibrahim FS, Hady FM, Abdel-Gaied SM, Eid MR (2010) Influence of chemical reaction on heat and mass transfer of non-Newtonian fluid with yield stress by free convection from vertical surface in porous medium considering Soret effect. *Appl. Math. Mech. - Engl. Ed.* 31(6): 675–684.

[21] Yih KA (1997) The effect of uniform lateral mass flux effect on free convection about a vertical cone embedded in a saturated porous medium. *Int. Comm. Heat Mass Transfer* 24(8): 1195–1205.

[22] Yih KA (1998) Uniform lateral mass flux effect on natural convection of non-Newtonian fluids over a cone in porous media. *Int. Comm. Heat Mass Transfer* 25(7): 959–968.

[23] Yih KA (1999) Effect of radiation on natural convection about a truncated cone. *Int. J. Heat Mass Transfer* 42: 4299 – 4305.

[24] Yih KA (1999) Coupled heat and mass transfer by free convection over a truncated cone in porous media: VWT/VWC or VHF/VMF. *Acta Mech.* 137 : 83–97.

[25] Murthy PVSN, Singh P (2000) Thermal dispersion effects on non-Darcy convection over a cone. *Compu. Math. Applications* 40: 1433 – 1444.

[26] Roy S, Anilkumar D (2004) Unsteady mixed convection from a rotating cone in a rotating fluid due to the combined effects of thermal and mass diffusion. *Int. J. Heat Mass Transfer* 47:1673–1684.

[27] Takhar HS, Chamkha AJ, Nath G (2004) Effect of thermophysical quantities on the natural convection flow of gases over a vertical cone. *Int. J. Eng. Sci.* 42: 243–256.

[28] Singh PJ, Roy S (2007) Unsteady mixed convection flow over a vertical cone due to impulsive motion. *Int. J. Heat Mass Transfer* 50: 949–959.

[29] Kumari M, Nath G (2009) Natural convection from a vertical cone in a porous medium due to the combined effects of heat and mass diffusion with non-uniform wall temperature/concentration or heat/mass flux and suction/injection. *Int. J. Heat Mass Transfer* 52: 3064–3069.

[30] Cheng C-Y (2009) Natural convection heat transfer of non-Newtonian fluids in porous media from a vertical cone under mixed thermal boundary conditions. *Int. Comm. Heat Mass Transfer* 36: 693 – 697.

[31] Cheng C-Y (2009) Soret and Dufour effects on natural convection heat and mass transfer from a vertical cone in a porous medium, *Int. Comm. Heat Mass Transfer* 36 : 1020 – 1024.

[32] Cheng C-Y (2010) Soret and Dufour effects on heat and mass transfer by natural convection from a vertical truncated cone in a fluid-saturated porous medium with variable wall temperature and concentration. *Int. Comm. Heat Mass Transfer* 37 : 1031 – 1035.

[33] Cheng C-Y (2010) Soret and Dufour effects on heat and mass transfer by natural convection from a vertical truncated cone in a fluid-saturated porous medium with variable wall temperature and concentration. *Int. Comm. Heat Mass Transfer* 37 : 1031 – 1035.

[34] Cheng C-Y (2010) Nonsimilar boundary layer analysis of double-diffusive convection from a vertical truncated cone in a porous medium with variable viscosity. *Int. Comm. Heat Mass Transfer* 37: 1031 – 1035.

[35] Choi SUS (1995) Enhancing thermal conductivity of fluid with nanoparticles. developments and applications of non-Newtonian flow. *ASME FED 231/MD* 66 : 99–105.

[36] Khanafer K, Vafai K, Lightstone M (2003) Buoyancy-driven heat transfer enhancement in a two-dimensional enclosure utilizing nanofluids. *Int. J. Heat Mass Transfer* 46 : 3639–3653.

[37] Buongiorno J (2006) Convective transport in nanofluids. *ASME J. Heat Transfer* 128 : 240–250.

[38] Daungthongsuk W, Wongwises S (2007) A critical review of convective heat transfer nanofluids. *Ren. Sustainable Energy Rev.* 11: 797–817.

[39] Oztop HF, Abu-Nada E (2008) Numerical study of natural convection in partially heated rectangular enclosures filled with nanofluids. *Int. J. Heat Fluid Flow* 29: 1326–1336.

[40] Nield DA, Kuznetsov AV (2009) The Cheng–Minkowycz problem for natural convective boundary-layer flow in a porous medium saturated by a nanofluids. *Int. J. Heat Mass Transfer* 52: 5792–5795.

[41] Nield DA, Kuznetsov AV (2011) The Cheng–Minkowycz problem for the double-diffusive natural convective boundary-layer flow in a porous medium saturated by a nanofluids. *Int. J. Heat Mass Transfer* 54: 374–378.

[42] Ahmad S, Pop I (2010) Mixed convection boundary layer flow from a vertical flat plate embedded in a porous medium filled with nanofluids. *Int. Comm. Heat Mass Transfer* 37: 987 - 991.

[43] Khan WA, Pop I (2010) Boundary-layer flow of a nanofluid past a stretching sheet. *Int. J. Heat Mass Transfer* 53: 2477–2483.

[44] Kuznetsov AV, Nield DA (2010) Natural convective boundary-layer flow of a nanofluid past a vertical plate. *Int. J. Thermal Sci.* 49: 243 - 247.

[45] Kuznetsov AV, Nield DA (2010) Effect of local thermal non-equilibrium on the onset of convection in a porous medium layer saturated by a nanofluid. *Transp. Porous Media* 83: 425–436.

[46] Bachok N, Ishak A, Pop I (2010) Boundary-layer flow of nanofluids over a moving surface in a flowing fluid. *Int. J. Thermal Sci.* 49: 1663 - 1668.

Chapter 6

Forced Convective Heat Transfer and Fluid Flow Characteristics in Curved Ducts

Tilak T. Chandratilleke and Nima Nadim

Additional information is available at the end of the chapter

1. Introduction

Curved fluid flow passages are common in most technological systems involving fluid transport, heat exchange and thermal power generation. Some examples are: compact heat exchangers, steam boilers, gas turbines blades, rocket engine nozzle cooling and refrigeration. Such flows are subjected to centrifugal forces arising from continuous change in flow direction and, exhibit unique fluid and thermal characteristics that are vastly different to those within straight passages.

The centrifugal action induced by duct curvature imparts two key effects on the fluid flow. It produces a lateral fluid movement directed from inner duct wall towards the outer wall in the axial flow, thus causing a spiralling fluid motion through the duct. This lateral fluid movement is manifested by large counter-rotating pairs of vortices appearing in the duct cross section and is referred to as secondary flow. The centrifugal action also forms a radial fluid pressure gradient positively biased towards the outer duct wall. The lateral fluid circulation takes place adversely to the radial pressure field and is dampened by the viscous effects. The combined actions of the positive radial pressure gradient and the viscous forces lead to the formation of a stagnant fluid region near the outer wall. Beyond a certain critical axial flow rate, the radial pressure gradient would exceed the equilibrium condition in the stagnant fluid region at outer wall and triggers a localised flow circulation that forms additional pairs of vortices. This flow condition is known as Dean Instability [1] and the additional vortices are called Dean Vortices.

In his pioneering work, Dean [1] proposed the dimensionless Dean Number $K = \left[\dfrac{D_h}{R}\right]^{\frac{1}{2}} \mathrm{Re}$ for characterising the secondary flow behaviour. Moffat [2] and Eustice [3] have experimentally observed and verified the critical velocity requirements for Dean Instability, while Baylis [4] and Humphery et al. [5] affirmed the use of Dean Number in designating secondary flow behaviour. However, subsequent studies by Cheng et al. [6], Ghia and Shokhey [7], and Sugiyama et al, [8] showed that the duct aspect ratio and curvature ratio also significantly influence Dean Instability in curved rectangular ducts.

For rectangular ducts, Chandratilleke et al. [9,10,11] reported an extensive parametric study, capturing the profound influence from duct aspect ratio, curvature ratio and wall heat flux on the flow behaviour. Using an approach based on the stream function, they developed a two-dimensional model simplified by dynamic similarity assumption in axial direction. This simulation showed very good correlation to the data from their own experimental work [10,11]. The locations of Dean vortex formation within the flow was qualitatively identified through the use of intersecting stream function contours of zero potential. Comprehensive results were presented within the Dean Number range of $25 \leq K \leq 500$, aspect ratio at $1 \leq Ar \leq 8$ and Grashof number at $12.5 \leq Gr \leq 12500$. It was identified that the onset of Dean Instability would strongly be dependent on the duct aspect ratio wherein more Dean vortices are produced in ducts of high aspect ratio. The wall heat flux radically changed the flow patterns and showed a tendency to suppress Dean vortex formation. Subsequently, Yanase [12] and Fellouah et al. [13,14] confirmed and validated the findings by Chandratilleke et al. [11].

The stream function approach is clearly limited to two-dimensional flows and does not accurately applied for real flow situations. In developing improved models, Guo et al. [15] considered laminar incompressible flow and formulated a three-dimensional simulation to explore the interactive behaviour of geometrical and flow parameters on heat transfer and pressure drop. Using flow entropy for hydrothermal assessment, they reported the influence from Reynolds number and curvature ratio on the flow profile and Nusselt number. For curved rectangular ducts, Ko et al. [16] proposed to split the overall flow entropy into individual contributions from heat transfer and fluid viscous friction. The relative magnitudes of these components were appraised to determine the heat or viscous-biased irreversibility in the flow domain. This approach was adopted for achieving thermal optimisation in ducts through minimised overall entropy, and illustrated for enhancing forced convection in curved ducts with longitudinal fins under laminar and turbulent conditions [17,18,19]. However, the entropy approach did not warrant precise identification of Dean Instability.

Fellouah et al. [13,14] also have presented a useful three-dimensional model for rectangular curved ducts of aspect ratio within $0.5 \leq Ar \leq 12$ and curvature ratio at $5.5 \leq \gamma \leq 20$ and. Considering both water and air as working fluids, their results were was experimentally validated using a semi-circular duct. These tests also provided visualisation data on vortex formation at several Dean Numbers. For detecting Dean Instability, they suggested the use of the radial gradient of the axial fluid velocity. In this, a limiting value for the radial gradient of axial velocity is arbitrarily assigned as a triggering threshold for Dean Instability. This approach represents an early attempt to incorporate Dean vortex detection to a simulation process rather than a trail-and-error method as previously practised. In spite of the perceived benefits in computation, this approach cannot be rationalised because the radial gradient of axial velocity is remotely linked with vortex generation. This short fall is reflected when the model of Fellouah et al. [13] failed to detect Dean Instability that was clearly observed in some reported cases [10,11]. With improved capabilities, Chandratilleke et al. [20,21] have suggested other approaches that will be discussed later.

Most experimental and numerical analyses on curved ducts have been performed on rectangular duct geometries. By virtue of shape, such ducts have less wall interference on secondary vortex formation, making it relatively easier for numerical modelling and convenient for experimentation including flow visualisation. Ducts with elliptical and circular cross sections have received less research attention in spite of being a common geometry used in most technological systems. The curved duct flow behaviour in such geometries remains relatively unexplored.

Dong and Ebadian [22], and Silva et al. [23] have numerically simulated the flow through elliptical curved channels and observed that the stagnation regions are vastly different to those in rectangular ducts. Unlike in rectangular ducts, Dean Instability was seen to originate within the flow rather than at the outer duct wall. Papadopoulos and Hatzikonstantinou [24] considered elliptical curved ducts with internal fins and have numerically investigated the effects of fin height on friction factor and heat transfer. They concluded that the appearance of secondary vortices next to the concave wall would make the friction factor dependent on both fin and duct heights. Andrade et al. [25] reported a study with a finite element numerical model and discussed the influence from temperature-dependent viscosity on heat transfer and velocity profile for fully-developed forced convection in elliptical curved tube. They considered both cooling and heating cases wherein the Nusselt Number was found to be lower for cases of variable viscosity than constant properties. This was attributed to the increased viscosity at the cooler inner duct wall dampening the secondary flow and the vortex formation.

The above review of published literature indicates that the available modelling methods have yet to develop for realistic representation of complex secondary flow behaviour and improved predictability for Dean Instability. Parametric influences of duct geometry and flow variables remain unexplored and poorly understood with no decisive approach for defining the onset of Dean vortices and their locations. Significantly improving these shortfalls, this chapter describes an advanced three-dimensional numerical simulation methodology based on helicity, which is congruent with the vortex motion of secondary flow. Facilitating much compliant tracking of vortex flow paths, the model uses a curvilinear mesh that is more effective in capturing the intricate details of vortices and flexibly applied to both rectangular and elliptical ducts. Two intuitive approaches for detecting the onset of Dean Instability are proposed and examined in the study. An extensive parametric investigation is presented with physical interpretation of results for improved understanding of flow behaviour. A thermal optimisation scheme based on flow irreversibilities is developed for curved ducts.

2. Nomenclature

Ar	Aspect Ratio= a/b
a	Height of cross section (mm)
b	Width of cross section (mm)
D_h	Hydraulic diameter = $\dfrac{ab}{\sqrt{\dfrac{(a+b)}{2}}}$ (mm)
F_c	Centrifugal force (N)
g	Gravity (m/s²)
H	Helicity (m/s²)
H^*	Dimensionless Helicity = $\dfrac{HD_h}{U_{in}^2}$
$\hat{i},\hat{j},\hat{k}$	Unit vectors in x,y,z directions
K	Dean number = $\left(\dfrac{D_h}{R}\right)^{\frac{1}{2}} Re$
p	Static pressure (Pa)
p^*	Dimensionless static pressure = $\dfrac{p}{\dfrac{1}{2}\rho U_{in}^2}$
R	Radius of curved channel (m)
Re	Reynolds number = $\dfrac{U_{in}D_h}{\upsilon}$
$\hat{S}$	Coordinate along duct cross section for defining secondary flow direction
U_{in}	Velocity at duct inlet (m/s)
u,v,w	Velocities component (m/s)
u^*,v^*,w^*	Dimensionless velocity = $\dfrac{u,v,w}{U_{in}}$
V_r	Axial velocity (m/s)
x,y,z	Coordinates (m)
x^*,y^*,z^*	Dimensionless coordinates= $\dfrac{x,y,z}{D_h}$

Greek symbols

γ	Curvature ratio= R/b
θ	Angular position of cross section (deg)
Φ	Dissipation function
ν	Kinematic Viscosity (m²/s)
μ	Dynamic viscosity (Ns/m²)
ρ	Density (kg/m³)
ω	

3. Model description and numerical analysis

Fig. 1 shows the rectangular and elliptical duct geometries used for the three-dimensional model developed in the current study. It also indicates the key geometrical parameters of duct height a, duct width b and the duct radius of curvature R. The geometrical model consists of a semi-circular curved duct test section fitted with a straight inlet passage to ensure fully-developed flow at entry to the curved duct and an outlet passage for smooth flow exit.

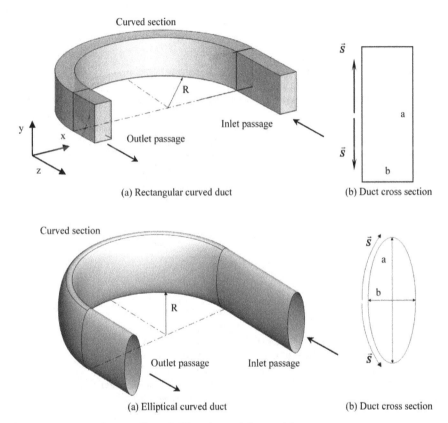

(a) Rectangular curved duct   (b) Duct cross section

(a) Elliptical curved duct   (b) Duct cross section

Figure 1. Geometry of rectangular and elliptical curved duct models

The analysis focuses on semi-circular curved ducts having fixed width b of 10 mm and constant radius of curvature R of 125 mm. The duct aspect ratio Ar is changed by varying duct height a. The working fluid air is assumed to be an incompressible Newtonian fluid with temperature-dependent fluid properties. The air enters at an inlet temperature of 300 K and flows steadily through the passage under laminar flow conditions. A uniform heat flux is applied on the outer wall (rectangular duct) or outer half of duct periphery (elliptical duct), while all the other walls being adiabatic. Constant velocity condition is applied to the inlet of the straight duct preceding the curved duct. The flow exit is taken to be a pressure outlet. The duct walls are assumed to have no slip boundary condition.

The numerical model solves the following governing equations:

Time-averaged continuity equation:

$$\nabla.(\vec{V}) = 0 \tag{1}$$

Momentum and Energy conservation equations:

$$\vec{V}.\nabla(\rho\vec{V}) = -\nabla p + \mu\nabla^2\vec{V} + \rho_m\vec{g} + \vec{F}_c \quad \text{(a)}$$
$$\vec{V}.\nabla(\rho c_p T) = k\nabla^2 T + S_T \quad \text{(b)} \tag{2}$$

The magnitude of the centrifugal body force term in radial direction is given by,

$$F_c = \rho\frac{V_r^2}{r} = \rho\frac{(u^2 + w^2)}{\sqrt{x^2 + z^2}} \tag{3}$$

Considering the position and alignment of curved part of geometry, centrifugal source term in Cartesian coordinate system is obtained from,

$$\vec{F}_c = \rho\frac{1 + sign(-z)}{2}\frac{(u^2 + w^2)}{x^2 + z^2}(x\hat{i} + z\hat{k}) \tag{4}$$

In Eq. 4, a Sign Function is used to ensure the centrifugal source term is applied only on the curved side of the geometry (i.e. $z \leq 0$). For obtaining the dimensionless parameters, the characteristics length, velocity and pressure are chosen to be D_h, U_{in}, $\frac{1}{2}\rho U_{in}^2$, respectively.

The thermally-induced buoyancy is included in the model by relating the density ρ_m in Eq. 2(a) to local fluid temperature. For this, a sixth-order polynomial as given by Eq. 5 is developed, where the coefficients are obtained using the published data. This approach is necessary because the linearity of the Boussinesq approximation caused some discrepancy and is found to be inadequate for evaluating the wall pressure gradient.

$$\rho(T) = 10^{-15}T^6 - 3\times10^{-12}T^5 + 3\times10^{-9}T^4 - 2\times10^{-6}T^3 + 6\times10^{-4}T^4 - 0.1008T + 9.3618 \tag{5}$$

In capturing the helix-like fluid motion of secondary flow, this three-dimensional model incorporates a helicity function, which is defined by Eq. 6. The helicity is non-dimensionalised using the reference helicity based on hydraulic diameter, as given by Eq. 7.

$$H = \vec{V}.\vec{\omega} = u(\frac{\partial w}{\partial y} - \frac{\partial v}{\partial z}) + v(\frac{\partial u}{\partial z} - \frac{\partial w}{\partial x}) + w(\frac{\partial v}{\partial x} - \frac{\partial u}{\partial y}) \tag{6}$$

$$H \approx \frac{U_{in}^2}{D_h} \Rightarrow H^* = H(\frac{D_h}{U_{in}^2}) \qquad (7)$$

For identifying the onset of Dean Instability, the model proposes two separate criteria. The first criterion is based on helicity, which is computed from Eq. 7. The second criterion uses the outer duct wall pressure gradient profile for which the non-dimensional wall pressure gradient is obtained from Eq. 8. A sign convention is incorporated in Eq. 8 to designate the opposite rotational directions of the vortices in upper and lower half of duct cross section, following the selection of coordinate $\vec{S}$ along the outer wall of duct cross section, as depicted in Fig. 1(b).

$$\frac{dp^*}{d\vec{S}} = \begin{cases} -\dfrac{D_h}{\frac{1}{2}\rho U_{in}^2} \dfrac{dp}{ds} & \text{at upper half of cross section} \\[2ex] \dfrac{D_h}{\frac{1}{2}\rho U_{in}^2} \dfrac{dp}{ds} & \text{at lower half of cross section} \end{cases} \qquad (8)$$

For the forced convective heat transfer in the duct, the local Nusselt Number is defined as $Nu = \dfrac{hD}{k}$ where the heat transfer coefficient h, is determined by considering the grid cell temperature difference between the heated wall and the adjacent fluid cell. The average Nusselt number is obtained from the surface integral, $\bar{Nu} = \dfrac{\int_A Nu dA}{A}$.

The model formulates a thermal optimisation scheme using flow irreversibility as a criterion. For this, the overall entropy generation is split in to the components of irreversibility contributed by the wall heat transfer, including thermally-induced buoyancy effects and, that due to fluid flow friction compounded by secondary fluid motion. As such, the components of entropy generation from heat transfer (S_T) and that from fluid friction (S_p) within the solution domain are evaluated using the expressions 9 (a), (b) and (c).

$$S_T = \frac{k}{T^2}\left[(\frac{\partial T}{\partial x})^2 + (\frac{\partial T}{\partial y})^2 + (\frac{\partial T}{\partial z})^2 \right] \qquad (a)$$

$$S_p = \frac{\mu}{T}\Phi \qquad (b)$$

$$(9)$$

where

$$\Phi = 2\,[(\frac{\partial u}{\partial x})^2 + (\frac{\partial v}{\partial y})^2 + (\frac{\partial w}{\partial z})^2] + (\frac{\partial u}{\partial y} + \frac{\partial v}{\partial x})^2 + (\frac{\partial w}{\partial y} + \frac{\partial v}{\partial z})^2 + (\frac{\partial u}{\partial z} + \frac{\partial w}{\partial x})^2 \qquad (c)$$

The entropy terms are generalised as volumetric-averaged values using,

$$S_T''' = \frac{\int S_T dv}{dv} \quad \text{(a)}$$

$$S_p''' = \frac{\int S_p dv}{dv} \quad \text{(b)} \quad\quad\quad (10)$$

$$S_g''' = S_T''' + S_p''' \quad \text{(c)}$$

Based on the above volumetric entropy generation terms, Bejan number is defined as,

$$Be = \frac{S_T'''}{S_g'''} \quad\quad\quad (11)$$

Bejan Number reflects the relative domination of flow irreversibility by heat transfer with respect to the overall irreversibility [26]. In this, magnitude 1.0 for Bejan Number indicates the entropy generation totally dominated by heat transfer while 0 refers to fluid friction causing all of the flow irreversibility. Thus for curved ducts, the Bejan number performs as a visual map signifying the relative strength of thermal effects within the fluid domain interacted by the secondary flow and Dean vortices. This approach is then utilised for thermally optimising the forced convection process in curved ducts.

Incorporating above governing equations, a finite volume-based CFD model is formulated with the commercial package FLUENT where SIMPLE algorithm is used for pressure-velocity coupling. The momentum and energy equations are discretised by first and second order schemes, respectively. Since the model considers both buoyancy and centrifugal source terms, pressure discretisation is performed with body force-weighted approach. The stability of the solution is monitored through continuity, velocity, energy and dimensionless helicty where the convergence is achieved with values not higher than 10^{-5}. Grid independency is carefully checked paying a special attention to boundary layer modelling at the outer wall.

The simulation is performed to obtain the profiles of velocity, helicity, temperature and Bejan Number at curved duct for a range of flow rates giving Dean Number within 80 to 1600. The wall heat flux is varied up to 1000 W/m^2. The results showing some signs of Dean vortices are further refined by re-running the simulations with closer steps of K to determine the exact point of instability and the critical Dean number. This procedure is repeated for all aspect ratios, flow rates and heat fluxes. Forced convection is evaluated by the local and average Nusselt numbers at the duct wall.

3.1. Mesh generation, grid sensitivity and model validation

In secondary flow simulations, the solution convergence is critically dependant on the grid selection because of the intricate flow patterns and intense flow gradients. For rectangular ducts, a fully structured mesh would be adequate. However, due to the geometrical shape, elliptical ducts tend to have much extreme flow field gradients and would require a more stringent grid arrangement.

In rectangular ducts a progressively reducing mesh is used with a much finer mesh near the outer wall where the onset of Dean Instability is anticipated. This approach has not been attempted in previous studies [11-13, 18-19] wherein it was argued that a mesh size less than 1 mm would not improve the accuracy, but only increase the computational time. The mesh refinement of the present analysis clearly identifies that a finer mesh near the wall is critical for detecting the onset of Dean vortices as accurately as possible. For testing grid dependency, the current study uses five mesh schemes indicated in Table 1. In this, columns A, B and C represent the number of grids over duct width, height and length, respectively, while the column D indicates the progressive reduction of grid size over duct width towards the outer wall.

Scheme	Number of Grids			
	A	B	C	D
Mesh1	26	51	305	1
Mesh2	31	64	305	1
Mesh3	43	84	305	1
Mesh4	50	98	305	1
Mesh5	26	51	305	1.05

Table 1. Grid selection and mesh schemes for rectangular ducts

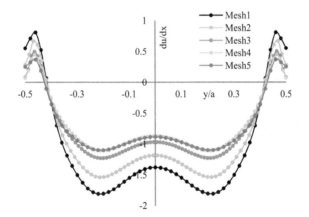

Figure 2. Grid dependancy test for rectangular duct using velocity derivative in y-direction at exit

$$K=130, Ar = 2$$

For rectangular ducts, Fig. 2 illustrates the grid dependency test conducted using the velocity derivative in y-direction at the duct outer wall. The Schemes 4 and 5 having mesh size less than 1 mm show better suitability than the other three schemes. However, the Scheme 5 is chosen as the optimum because of its slightly larger cell volume arising from progressively varied mesh size. This approach remarkably improved the vortex capturing ability in the solution domain without excessive computational demand. As such, the present study performed all computations with Scheme 5 of mesh size less than 1 mm, achieving much higher accuracy than any previously reported work.

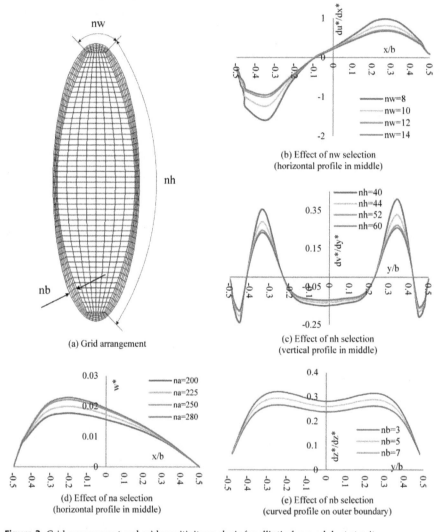

Figure 3. Grid arrangement and grid sensitivity analysis for elliptical curved duct at exit

For handling steeper flow gradients in elliptical ducts, the current analysis uses the different grid arrangement depicted in Fig. 3. This scheme divides the elliptical duct cross section into five regions, as shown in Fig. 3(a) and, the rectangular cells are swept along the axial direction to obtain a fully hexagonal structured mesh. The grid distribution is determined by the parameters nw, nh, nb and na, which are respectively, the number of grids on duct width, duct height and duct periphery while *na* represents the number of layers along the axial direction. This formation ensures uniform grid residual throughout the solution domain. Figs. 3(b), (c) and (d) show the grid sensitivity analysis in terms of nw, nh and nb and na, where 36 permutations are considered. For elliptical duct simulation, the optimal grid selection is taken to be nw = 12, nh = 52, nb = 7 and na = 250.

In validating results for a rectangular duct, Figs. 4(a) and 4(b) show a comparison of the axial flow velocity in x and y directions from the current model with those from Ghia and Shokhey [7] and Fellouah et al. [14]. It is seen that both magnitudes and trends of axial velocity are very favourably matched confirming the integrity of the current numerical process. Similarly for an elliptical duct, Figs. 4(c) and 4(d) provide a comparison of the axial velocity profiles at the mid cross section planes from the current study with those from Dong and Ebadian [22] and Silva et al. [23]. A very good agreement is clearly evident, validating the numerical consistency of the present simulation.

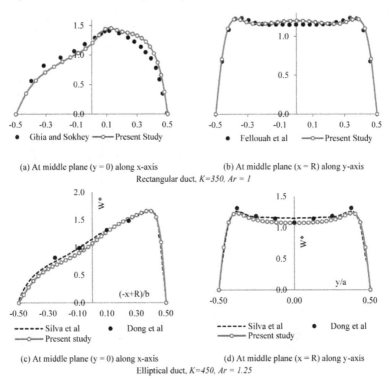

(a) At middle plane (y = 0) along x-axis (b) At middle plane (x = R) along y-axis

Rectangular duct, $K=350$, $Ar = 1$

(c) At middle plane (y = 0) along x-axis (d) At middle plane (x = R) along y-axis

Elliptical duct, $K=450$, $Ar = 1.25$

Figure 4. Model validation using dimensionless axial velocity profile at curved duct exit

4. Results and discussion

4.1. Fluid flow characteristics and geometrical influence

Fig. 5 depicts typical dimensionless helicity profiles at the exit of both rectangular and elliptical curved ducts for several Dean numbers with no external wall heating. These patterns show unique flow features that are not present in straight ducts. Initially at low K, the flow profiles indicate two large counter-rotating vortices that are attributable to the centrifugal action from the duct curvature.

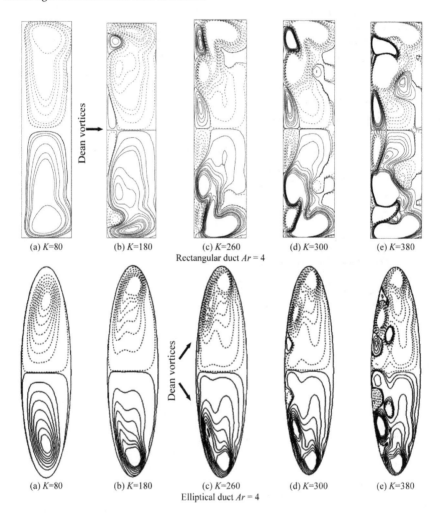

(a) $K=80$ (b) $K=180$ (c) $K=260$ (d) $K=300$ (e) $K=380$

Rectangular duct $Ar = 4$

(a) $K=80$ (b) $K=180$ (c) $K=260$ (d) $K=300$ (e) $K=380$

Elliptical duct $Ar = 4$

Figure 5. Dimensionless helicity contours in curved ducts for varied flow rates-no wall heating

In curved ducts, the centrifugal action is manifested as two key effects. It generates a positive radial pressure field directed towards the outer duct wall (left wall in Fig. 5). Within this positive (adverse) pressure field, the centrifugal force drives the fluid radially away from the inner wall towards the outer duct wall. This sets up a lateral fluid circulation called secondary flow, which is manifested as the formation of counter-rotating vortices observed in Fig 5 in the duct cross section.

Increased axial flow rate (larger K) makes the secondary fluid motion more vigorous to produce stronger vortices. The positive radial pressure field is also Higher flow rate also intensifies that imparts more adverse effect on the secondary fluid motion. This latter effect adversely acts on to slow it down assisted by fluid viscosity. Consequently near the outer duct wall, a stagnant flow region is formed and a state of fluid equilibrium is established. Above a certain critical value of K, the radial pressure field becomes far greater for fluid equilibrium within the stagnant region and a weak local fluid recirculation is triggered. This flow situation is referred to as the Dean Instability identified by the critical value of the Dean number. The local fluid circulation triggering flow instability is manifested as pairs of additional vortices called Dean Vortices. In Fig. 5, the onset of Dean Instability occurs within $K = 80$ to 180 for the rectangular duct while for the Dean vortices between $K = 180$ to 260. These Dean vortices are designated as the helicity contour corresponding to $H^* = 0.01$. Upon inception, Dean vortices gradually grow bigger with increasing K, as depicted.

It is noticed that in rectangular ducts, the Dean instability tends to occur at a lower K than in elliptical ducts of the same aspect ratio. This is because in rectangular ducts, the secondary vortex motion is less steeply deflected at the outer wall by the cross sectional geometry than in elliptical ducts allowing more freedom for fluid movement and Dean vortex formation.

4.2. Dean instability and detection of Dean vortices

In curved ducts, the appearance of Dean vortices is traditionally identified through tedious flow visualisation or by numerical trial-and-error approach. In the latter, simulations are repeatedly performed to gradually narrow down the range of K for determining the critical Dean number and the flow patterns within a chosen tolerance limit. This involves guesswork and significant computational time. Chandratilleke et.al. [11] successfully developed a criterion based on zero-potential stream function contours to identify the locations of Dean vortex generation. Whilst that approach is adequate for two-dimensional cases, it is not applicable for three-dimensional flow. The work of Fellouah et al. [13] used the radial gradient of the axial velocity as a measure of identifying the flow instability. Such technique is not justifiable because the axial velocity change in radial direction is not physically connected with the secondary vortex generation. This inadequacy is clearly reflected in the work of Fellouah et al. [13], where their simulation failed to detect Dean vortices for some basic flow conditions. For example around $K = 180$ for a rectangular duct, Fellouah et al. [13] and Silva et al. [23] did not detect Dean vortices while the current study clearly predicts such vortices.

As of now, a reliable technique for identifying Dean vortices is not available in literature, particularly for three-dimensional simulations. The formulation of a generalised approach is difficult because the inception of Dean vortices are influenced by strongly inter-dependant effects of duct geometry and the flow variables. In forming a technique for Dean vortex detection, the present study proposes two practical criteria that can be integrated into and directly performed within the computational process.

4.2.1. Criterion 1 - Helicity threshold method

This criterion utilises dimensionless helicity H^* for tracking the onset of Dean Instability. It assigns the minimum threshold value for H^* by which the helicity contours are demarcated in the flow domain. These contours are designated as Dean vortices. This selection of H^* threshold essentially depends on the contour detection accuracy required, similar to defining the boundary layer thickness with a chosen velocity tolerance in traditional fluid flows.

Through exhaustive simulation runs, the current study has identified and proposes $H^* = \pm$ 0.01 to be an excellent choice that warrants precise and consistent detection of Dean vortices for all cases examined. A key advantage of this technique is that the detection precision can be varied to suit the required accuracy. Also, the technique can be readily integrated into the computational process to perform locally in the solution domain rather than as a cumbersome post-processing method. Hence, the determination of Dean vortices is more precise and less time consuming. The achievable precision is demonstrated below in Fig. 6, where the helicity contours obtained by using $H^* = 0.01$ are shown just before and just after the onset of Dean vortices (indicated by arrows).

For the rectangular duct, the Dean vortices are absent in the flow for $K = 95$. However, when K is increased to 102, the helicity contours indicating Dean vortices are first detected. Similarly for the elliptical duct, the appearance of Dean vortices is detectable within $K = 230$ to 235. By adopting H^* threshold less than 0.01, this range can be further narrowed to improve precision of K value that corresponds to the onset of Dean Instability. The critical Dean numbers thus determined are: $K = 100$ for the rectangular duct and $K = 234$ for the elliptical duct. However, evidently such refinements to H^* would only incur a marginal benefit towards precision at the expense of significant increase in computational demand. Therefore, $H^* = 0.01$ is concluded to be a very appropriate threshold for detecting Dean Instability. Previous methods never provided this degree of accuracy, flexibility or ability for vortex detection, signifying that this method is far superior to any reported approach including that by Ghia and Shokhey [7] and Fellouah et al. [13] using the axial velocity gradient.

4.2.2. Criterion 2 - Adverse wall pressure gradient method

This criterion is based on the unique features of the fluid pressure distribution along the outer duct wall. The gradient of this pressure profile shows inflection points that initially remain negative at low flow velocities and gradually shifts towards positive magnitudes as the flow rate is increased. Outer wall duct locations indicating negative-to-positive gradient change strongly correlates to the appearance of Dean vortices and forms the basis for a criterion to identify Dean Instability. These characteristics are illustrated in Fig. 7.

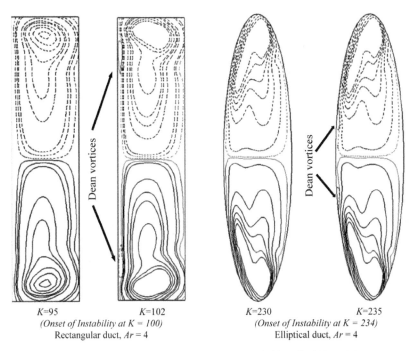

$K=95$ $K=102$ $K=230$ $K=235$
(Onset of Instability at K = 100)   *(Onset of Instability at K = 234)*
Rectangular duct, $Ar = 4$ Elliptical duct, $Ar = 4$

Figure 6. Detection of Dean vortex formation using helicity threshold method $H^* = \pm 0.01$

For several Dean Numbers, Fig. 7 shows typical outer wall pressure gradient profiles evaluated from Eq. 8 for a rectangular duct in Fig. 7(a) and, for elliptical ducts in Figs. 7 (b) and (c). The variable S is the displacement coordinate along the outer duct wall boundary, as indicated in Fig. 1. At low values of K in Fig. 7(a), the pressure gradient remains negative over the entire outer wall. As K increases, the inflection points of the profile gradually shift to towards a positive magnitude, thus creating regions of adverse pressure gradients at the outer wall. The corresponding helicity contours indicate that these localities at the outer wall would develop flow reversal leading to the generation of Dean vortices. At $K = 139$, the inflection points have just become positive, representing the critical Dean Number and the onset of Dean Instability.

Showing similar trends, Fig. 7(b) illustrates the pressure gradient profile for an elliptical duct of aspect ratio 3. At $K = 285$, the entire pressure profile remains negative over the outer wall and the corresponding helicity patterns in Fig. 8(a) show Dean vortices are absent in the flow. At $K = 291$, the pressure gradient just changes from negative to positive at the duct centre and the first appearance (onset) of Dean vortices is noted in the flow profile of Fig. 8(a). For $K = 297$, the pressure gradient profile shows distinctly positive regions that well correlate to Dean vortex locations, as depicted.

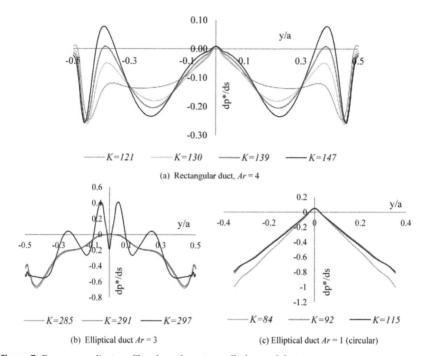

(a) Rectangular duct, $Ar = 4$

(b) Elliptical duct $Ar = 3$ (c) Elliptical duct $Ar = 1$ (circular)

Figure 7. Pressure gradient profiles along the outer wall of curved duct

Above observations conclusively indicate that the adverse (positive) pressure gradients at the outer duct wall could be effectively used for detecting the onset and location of Dean vortices. However, this approach has some drawback as discussed below.

Fig. 7(c) shows the outer wall pressure gradient for an elliptical duct of aspect ratio $Ar = 1$ (circular). This profile does not exhibit the negative-to-positive changeover in the pressure gradient as for elliptical ducts with Ar > 1 illustrated in Figs. 7(a) and 7(b), meaning that Dean vortices are not predicted according to the adverse pressure criterion. However to the contrary, Fig. 8(c) indeed shows the onset of Dean vortices at $K = 92$. This may seem a contradiction that is clarified below.

For a duct with $Ar = 1$ (circular duct), Fig. 8(c) clearly shows that the Dean vortices appear in the centre of the flow, but not at the outer wall as with rectangular and elliptical ducts. Therefore, the outer wall pressure gradient has lesser bearing on the flow reversal associated with Dean vortex formation. As such, the pressure gradient criterion tends to over predict the critical Dean Number for instability and has diminished suitability for elliptical ducts with aspect ratio near unity. It would perform satisfactorily for ducts with aspect ratio Ar > 1 where Dean vortices are formed at outer duct wall. Upon this overarching scrutiny, the helicity threshold method can be regarded as a precise, reliable and universally applicable technique for detecting Dean vortices in curved ducts of any shape and aspect ratio.

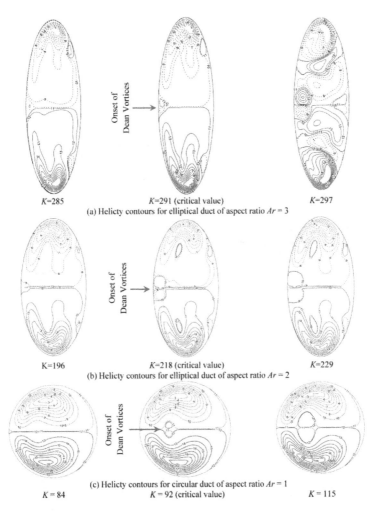

Figure 8. Dean vortex formation in elliptical ducts of different aspect ratios *(detected using H* = 0.01)*

4.2.3. Effect of duct aspect ratio

Obtained by using the helicity threshold method in the current study, Fig. 9 illustrates the influence of duct aspect ratio on the critical flow requirement for triggering Dean Instability in both rectangular and elliptical ducts. For comparison, the figure also provides the critical Dean number from the experimental results of Chen et al. [6] for rectangular ducts.

It is noted that the critical Dean number initially increases with the aspect ratio up to a certain value and then falls away for higher K. This behaviour conforms to the trend shown in Fig. 8 for elliptical ducts and the observations reported in previous experimental and numerical work. For rectangular ducts, the current model under-predicts the critical requirement for Dean Instability compared to the experimental results of [6]. This is because, the helicity method identifies Dean vortices much early in the growth process, whilst in experiments, the vortices are visually observed after they have grown to a detectable size at a higher Dean number.

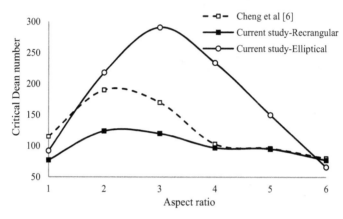

Figure 9. Effect of aspect ratio on critical Dean number

4.3. Thermal characteristics and forced convection

4.3.1. Effects of flow rate

The helicity contours and the fluid temperature fields in Fig. 10 illustrate the effect of outer wall heating in rectangular and elliptical ducts for two selected values of K. The comparison of helicity contours in Figs. 10 and 5 indicates the wall heating essentially sets up a convective fluid circulation that interacts with the secondary flow and significantly alters the flow characteristics. This fluid circulation is driven by the buoyancy forces resulting from thermally-induced density changes.

As evident in Fig. 10 for K = 80, at low flow rates, the convective circulation dominates the flow behaviour with less influence from secondary vortices. When the flow rate is increased, the centrifugal action intensifies and more vigorous secondary flow develops in the duct. Consequently, the secondary vortices overcome the thermal buoyancy effects to become more dominant and determine the overall fluid behaviour, as illustrated in Fig. 10 for K = 380.

Due to the confined geometry, the fluid flow within elliptical ducts is generally more constrained and tends to have steeper fluid flow gradients than in rectangular ducts. This geometrical effect is more pronounced at low flow rate when the flow patterns are dominated by thermally-induced buoyancy. Fig. 10 with K = 80, clearly demonstrates that the elliptical duct has sharper velocity and temperature gradients at the outer wall compared to the rectangular duct. Consequently, the elliptical duct exhibits a higher rate of forced convection than in the rectangular duct within the lower range of K in Fig. 11. For increased flow rate, the secondary vortices begin to dominate the flow behaviour. As shown in Fig. 10 for $K = 380$, the rectangular duct has steeper velocity and temperature gradients at the outer wall compared to elliptical duct. This gives rise to a higher forced convection in rectangular ducts compared to the elliptical duct when K approximately exceeds 350, as depicted in Fig. 11.

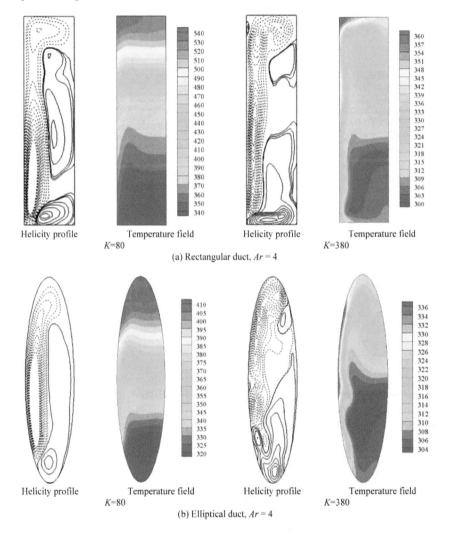

Figure 10. Outer wall heating effect on helicity and temperature profiles, $q = 250$ W/m^2

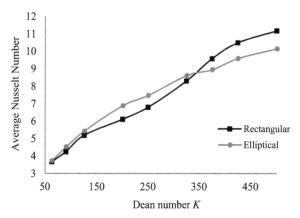

Figure 11. Fig. 11 Variation of average Nusselt number at outer wall with Dean number, $Ar = 4$, $q = 100$ W/m^2

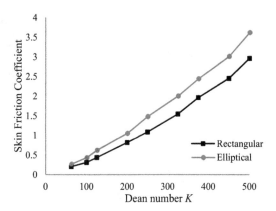

Figure 12. Variation of duct skin friction factor with Dean number, $Ar = 4$, $q = 100$ W/m^2

For elliptical and rectangular ducts, Fig. 12 shows the comparison of skin friction factor, which essentially depends on the velocity gradient and is related to the duct pressure drop. With inherently steep fluid profiles, the elliptical duct has slightly larger skin friction compared to rectangular ducts with no crossover in the entire flow range.

4.3.2. Effects of outer wall heating

Comparison of helicity contours in Figs. 10 and 5 at $K = 380$ indicates that with outer wall heating, a tendency is developed to impede the formation process of Dean vortices in both types of ducts. This is instigated by the thermal buoyancy-driven convection that continually acts to displace the fluid layer at the outer wall, thus weakening the formation of the stagnant fluid region, where the Dean Instability would occur. Hence, the flow reversal is undermined with diminished potential for Dean vortex generation. High wall heat fluxes would impart more adversity on the triggering flow conditions of Dean Instability. In Fig. 13, these wall heating effects are clearly evident as a gradual decline in Nusselt number with increased outer wall heat flux for both types of ducts.

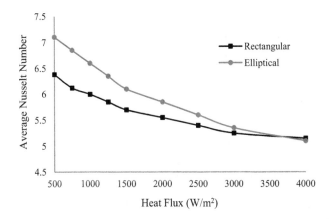

Figure 13. Variation of average Nusselt number at outer wall with outer wall heat flux, $Ar = 4$, $K = 250$

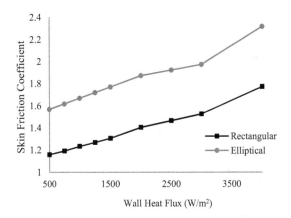

Figure 14. Variation of average duct Skin Friction with outer wall heat flux, $Ar = 4$, $K = 250$

With high wall heat fluxes, thermally induced convective circulation becomes stronger, imparting additional resistance to the axial fluid motion. This effect is more pronounced in elliptical ducts than in rectangular ducts. Consequently, duct skin friction coefficient tends to be relatively higher for elliptical ducts while indicating some increase with the wall heat flux, as evident in Fig. 14.

4.4. Thermal optimisation

Fig. 15 shows the Bejan number (Be) distribution in curved duct cross sections. For both rectangular and elliptical ducts at lower K, the flow irreversibility is practically dominated by the entropy generation from wall heat transfer (Be $\approx$ 1 with a red cast) over the entire flow cross section. As K increases, Bejan Number contours gradually acquire magnitudes less than 1 indicated by the blue cast. This signifies that in curved ducts, the secondary flow provides favourable fluid mixing to transport the hot fluid away from the heated wall, which in turn will improve forced convection. Magnitudes of Be < 1 also signifies an increased contribution to the total irreversibility from viscous effects, which would negate the overall flow benefits. These effects are less pronounced in rectangular ducts than in elliptical ducts. The opposing thermal and hydrodynamic trends identify a potentially useful technique for thermal optimisation of fluid flow through heated curved ducts, as explained below.

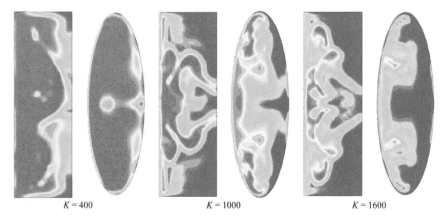

$K = 400$ $K = 1000$ $K = 1600$

Figure 15. Bejan Number contours for rectangular and elliptical curved ducts, $q = 100$ W/m²

Fig. 16 illustrates the variation of the flow irreversibility components S_t, S_p and S_g for rectangular and elliptical ducts with K, evaluated using Eq. 9 and 10. It is evident from the figure that, when K is increased, S_t steadily falls while S_p rapidly grows. Although this falling trend is similar to both types of ducts, the thermal irreversibility component S_t decays more gently in elliptical duct compared to the rectangular duct. This means, the elliptical duct is less efficient in transporting heat from the wall than the rectangular duct. On the other hand, the viscous irreversibility component S_p grows more rapidly in elliptical ducts, signifying that the elliptical shape causes higher frictional pressure loses.

The overall irreversibility S_g, which is the sum of S_t and S_p, is initially dominated by the thermal irreversibility component and steadily falls with K for both duct shapes. However, the falling gradient is steeper for the rectangular duct, which implies more effective heat transfer in the fluid domain with lesser adversity from viscous effects.

At a certain value of K, S_g reaches a minimum. This point of inflection is identified as practically the best possible "trade-off" between the highest achievable thermal benefits with the least viscous penalty for curved duct flows. It is noted that the magnitude of the lowest overall irreversibility is higher for elliptical ducts than for rectangular ducts, indicating rectangular ducts would thermally better perform.

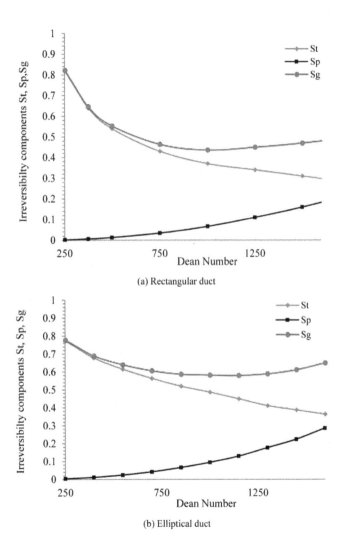

(a) Rectangular duct

(b) Elliptical duct

Figure 16. Curved duct thermal optimisation using total entropy generation

5. Conclusions

This book chapter provides a broad overview of the reported findings from the published numerical and experimental studies on fluid flow in curved duct geometries. It then presents and discusses the latest contributions to this field realised through an extensive research programmes of the authors where new analytical techniques and methodologies have been discovered for clearer fundamental understanding on fluid and thermal behaviour in curved ducts.

Improving accuracy over previous methods, the current study formulates a new three-dimensional simulation methodology based on helicity that realistically represents the secondary vortex structures in curved ducts of any shape. The model accurately identifies and predicts unique features of secondary flow and the associated forced convection in both rectangular and elliptical curved ducts. Conforming to the limited data in literature, the predicted results examine the effects of fluid flow rate, duct aspect ratio and heat flux over a wide practical range. The duct geometry and aspect ratio are found to have critical influence on the secondary flow characteristics, indicating a profound effect on Dean vortex formation, Dean Instability and forced convection. Elliptical ducts show marginally inferior thermal performance compared to rectangular ducts under identical flow conditions.

Overcoming a current analytical limitation, the study appraises two novel methods for detecting the onset of Dean Instability in curved ducts, ascertaining their technical feasibility. One such approach uses a defined helicity threshold while the other makes use of features of the outer wall pressure gradient. Both methods offer the process flexibility by integrating into and performed within a computational scheme, for fast, reliable and accurate detection of Dean vortex formation. However, the helicity threshold method is recognised to be more universally applied to all duct shapes and aspect ratios. The pressure gradient approach shows limited functionality for it over-predicts the critical Dean Number in curved ducts of aspect ratio near unity.

The study also presents a useful thermal optimisation technique for heated curved duct based on the Second Law irreversibilities. It identifies the elliptical ducts to have less favourable overall thermal and hydrodynamics characteristics than rectangular ducts.

Author details

Tilak T. Chandratilleke[*] and Nima Nadim
Department of Mechanical Engineering, Curtin University, Perth, Australia

6. References

[1] Dean, W.R. (1928), Fluid motion in a curved channel, Proc. Roy. Soc. London Ser, A121: 402–420.

---

[*] Corresponding Author

[2]   Moffatt, H.K. (1964), Viscous and resistive eddies near a sharp corner, Journal of Fluid Mechanics, 48 (1): 1–18.

[3]   Eustice, J. (1929), Experiments of streamline motion in curved pipes, Proc. Roy. Soc. London Se, 85 (1).

[4]   Baylis, J.A. (1971), Experiments on laminar flow in curved channels of square section, Journal of Fluid Mechanics. 48 (3): 417–422.

[5]   Humphrey, J.A.C., Taylor, A.M.K., Whitelaw, J.H. (1977), Laminar flow in a square duct of strong curvature, Journal of Fluid Mechanics, 83: 509–527.

[6]   Cheng, K.C., Nakayama, J., Akiyama, M. (1977), Effect of finite and infinite aspect ratios on flow patterns in curved rectangular channels, in: Flow Visualization International Symposium, Tokyo, p. 181.

[7]   Ghia, K.N., Sokhey, J.S. (1977), Laminar incompressible viscous flow in curved ducts of rectangular cross-section, Trans. ASME I: Journal of Fluids Engineering, 99: 640–648.

[8]   Sugiyama, Hayashi, Taro, Yamazaki, Koji.(1983), Flow Characteristics in the Curved Rectangular Channels, Bullletin of JSME, 26 (216): 532-552.

[9]   Chandratilleke, T.T.(2001), Secondary flow characteristics and convective heat transfer in a curved rectangular duct with external heating, 5th World Conference On Experimental Heat Transfer, Fluid Mechanics and Thermodynamics, Thessaloniki, Greece.

[10] Chandratilleke, T.T.(1998), Performance enhancement of a heat exchanger using secondary flow effects, in: Proc. of 2nd Pacific—Asia Conf.Mech. Eng., Manila, Philippines.

[11] Chandratilleke, Tilak T., Nursubyakto.(2002), Numerical prediction of secondary flow and convective heat transfer in externally heated curved rectangular ducts, International Journal of Thermal sciences 42: 187–198.

[12] Yanase, S. R. N. M., Kaga, Y.(2005), Numerical study of non-isothermal flow with convective heat transfer in a curved rectangular duct, International Journal of Thermal Sciences 44: 1047–1060.

[13] Fellouah, H., Moctar, A. Ould El, Peerhossaini, H.(2006), A criterion for detection of the onset of Dean instability in Newtonian fluids, European Journal of Mechanics B/Fluids 25: 505–531.

[14] Fellouah, H., Castelain, Ould-El-Moctar, Peerhossaini.(2010), H., The Dean instability in power-law and Bingham fluids in a curved rectangular duct, Journal of Non-Newtonian Fluid Mech. 165: 163–173.

[15] Guo, Jiangfeng, Xu, Mingtian, Cheng, Lin. (2011), Second law analysis of curved rectangular channels, International Journal of Thermal Sciences 50: 760-768.

[16] Ko. T.H, Ting, K.(2006), Entropy generation and optimal analysis for laminar forced convection in curved rectangular ducts: A numerical study, International Journal of Thermal Sciences 45: 138–150.

[17] Ko. T. H.(2005), Numerical investigation on laminar forced convection and entropy generation in a curved rectangular duct with longitudinal ribs mounted on heated wall." International Journal of Thermal Sciences 45: 390–404.

[18] Ko. T. H.(2006), A numerical study on entropy generation and optimization for laminar forced convection in a rectangular curved duct with longitudinal ribs, International Journal of Thermal Sciences 45: 1113–1125.

[19] Ko. T.H, C.P. Wu. (2008), A numerical study on entropy generation induced by turbulent forced convection in curved rectangular ducts with various aspect ratios." International Communications in Heat and Mass Transfer, 36: 25-31.

[20] Chandratilleke, Tilak T, Nadim, Nima and Narayanaswamy, Ramesh. (2010), Secondary flow characteristics and prediction of Dean vortices in fluid flow through a curved duct, 6th Australasian Congress on Applied Mechanics, Perth, Western Australia, Australia.

[21] Chandratilleke, Tilak T, Nadim, Nima and Narayanaswamy, Ramesh. (2011), Entropy-based secondary flow characterisation and thermal optimisation of fluid flow through curved heated rectangular ducts, 9th Astralasian Heat and Mass Transfer Cconference – 9AHMTC, Melbourne, Victoria, Australia.

[22] Dong Z.F., Ebadian M.A.(1991), Numerical analysis of laminar flow in curved elliptic ducts, J. Fluid Eng. 113:555-562.

[23] Silva, Ricardo Junqueira, Valle, Ramon Molina, Ziviani, Marcio.(1999), Numerical hydrodynamic and thermal analysis of laminar flow in curved elliptic and rectangular ducts, International Journal of Thermal Sciences, 38: 585-594.

[24] Papadopoulos, P.K., Hatzikonstantinou, P.M.(2008), Numerical study of laminar fluid flow in a curved elliptic duct with internal fins, International Journal of Heat and Fluid Flow 29: 540–544.

[25] Andrade, Claudia R. and Zaparoli, Edson L.(2001), Effect of temperature-dependent viscosity on fully developed laminar forced convection in curved duct, International communication in heat and mass transfer, 28(2):211-220.

[26] Bejan, Adrian. (1996), Entropy Generation Optimization, CRC Press Inc.

Chapter 7

Forced Turbulent Heat Convection in a Rectangular Duct with Non-Uniform Wall Temperature

G.A. Rivas, E.C. Garcia and M. Assato

Additional information is available at the end of the chapter

1. Introduction

Rectangular ducts are widely used in heat transfer devices, for instance, in compact heat exchangers, gas turbine cooling systems, cooling channels in combustion chambers and nuclear reactors. Forced turbulent heat convection in a square or rectangular duct is one of the fundamental problems in the thermal science and fluid mechanics. Recently, Qin and Fletcher [1] showed that Prandtl's secondary flow of the second kind has a significant effect in the transport of heat and momentum, as revealed by the recent Large Eddy Simulation (LES) technique. Several experimental and numerical studies have been conducted on turbulent flow though of non-circular ducts: Nikuradse [2]; Gessner and Emery [3]; Gessner and Po [4]; Melling and Whitelaw [5]; Nakayama et al. [6]; Myon and Kobayashi [7]; Assato [8]; Assato and De Lemos [9]; Home et al. [10]; Luo et al. [11]; Ergin et al. [12] Launder and Ying [13]; Emery et al. [14]; Hirota et al.[15]; Rokni [16]; Hongxing [17]; Yang and Hwang [18]; Park [19];Zhang et al. [20]; Zheng et al. [21]; Su and Da Silva Neto [22]; Saidi and Sundén [23]; Rokni [24]; Valencia [25]; Sharatchandra and Rhode [26]; Campo et al. [27]; Rokni and Sundén [28]; Yang and Ebadian [29] and others. The Melling and Whitelaw's [5] experimental work shows characteristics of turbulent flow in a rectangular duct where they have been used a laser-Doppler anemometer in which report the axial development mean velocity, secondary mean velocity, etc. Nakayama et al. [6] show the analysis of the fully developed flow field in rectangular and trapezoidal cross-section ducts; finite difference method was implemented and the model of Launder and Ying [13] has been used. On the other hand, Hirota et al. [15] present an experimental work in turbulent heat transfer in square ducts; they show details of turbulent flows and temperature fields. Likewise, Rokni [16] carried out a comparison of four different turbulence models for predicting the turbulent Reynolds stresses, and three turbulent heat flux models for square ducts. The literature presents various turbulence modeling in which confirm Linear Eddy Viscosity Models (LEVM) can give inaccurate predictions for Reynolds normal stresses: it does not have ability to predict secondary flows in non-circular ducts due to its isotropic treatment. In spite of that, they is one of the most popular model in the engineering due to its simplicity, good numerical stability in which can be applied in a wide variety of flows. Thus, the Nonlinear Eddy Viscosity Model (NLEVM) represents a progress of the classical LEVM in which this last one gives inequality treatment of the Reynolds normal stresses, needs of conditions for calculating turbulence-driven secondary flow in non-circular ducts and it has relatively high cost for solving the necessary two-equation formulation. The Reynolds Stress Model (RSM), also called second order or second moment closure model, is very accurate in the calculation of mean flow properties and Reynolds stresses, for simple to more complex flows including wall jets, asymmetric channel, non-circular duct and curved flows. However, RSM has some disadvantages, such as, very large computing costs. For calculating the turbulent heat fluxes, the Simple Eddy Diffusivity (SED) and Generalized Gradient Diffusion Hypothesis (GGDH) models have been adopted and investigated. Most of the works presented in the literature show results assuming constant temperature on the wall. However, in many engineering applications the heat fluxes and surface temperatures are non-constants around the duct, therefore becoming important the knowledge of the variation of the conductance around the duct, according to Kays and Crawford [30]. According to Garcia's developments [31], it is possible to carry out analysis with non-constant wall temperature boundary conditions. In this case, it is necessary to define a value that represents the mean wall temperatures in a given cross section, in which he has named T_{wm}. Following this treatment, for the present paper, important results have been computed and they are here being presented for rectangular cross-section ducts. Fluids such as air and water were analyzed under the influence of non-constant wall temperature distributions, with consequent presentations of resulted about turbulent convective heat exchanges and flow temperature profiles.

2. Mathematical formulation

2.1. Governing equations

The Reynolds Averaged Navier Stokes (RANS) equation system is composed of: continuity equation (1), momentum equation (2), and energy equation (3).

$$\frac{\partial}{\partial x_j}(U_j) = 0 \tag{1}$$

$$\frac{\partial}{\partial x_j}(U_i U_j) = -\frac{1}{\rho}\frac{\partial P}{\partial x_i} + \frac{1}{\rho}\frac{\partial}{\partial x_j}\left[\mu\left(\frac{\partial U_i}{\partial x_j} + \frac{\partial U_j}{\partial x_i}\right)\right] + \frac{1}{\rho}\frac{\partial}{\partial x_j}(-\rho\overline{u_i u_j}) \tag{2}$$

$$\frac{\partial}{\partial x_j}(U_j T_f) = \frac{1}{\rho}\frac{\partial}{\partial x_j}\left[\frac{\mu}{Pr}\frac{\partial T_f}{\partial x_j} + \left(-\rho\overline{u_j t}\right)\right] \tag{3}$$

For analyses of fully developed turbulent flow and heat transfer, the following hypothesis has been adopted: steady state, condition of non-slip on the wall and fluid with constant properties. The turbulent Reynolds stress $(-\rho\overline{u_i' u_j'})$ and the turbulent heat flux $(-\rho\overline{u_j' t'})$ were modeled and solved by algebraic and/or differential expressions.

2.2. Turbulence models for reynolds stresses

2.2.1. Nonlinear Eddy Viscosity Model (NLEVM)

The NLEVM Model to reproduce the tensions of Reynolds, it is necessary to include non-linear terms in the basic constitutive equations. This is done by attempting to capture the sensitivity of the curvatures of the stream lines. This model is based on the initial proposal of Speziale [35]. The Reynolds average equations, Equations (1) to (3), are applied for the device presents in the Figure 1(a) and (b).

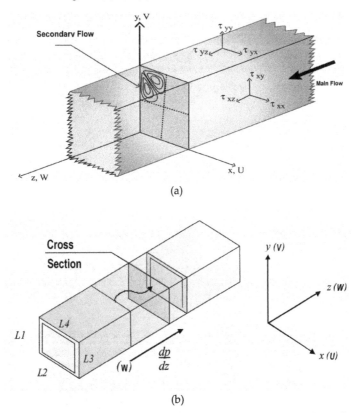

Figure 1. (a) Fully developed turbulent flows in rectangular ducts, (b) Rectangular duct: reference system and transversal section.

The velocity components U *and* V represent the secondary flow, and the axial velocity component W, the velocity of the main flow. The transport equations in tensorial form for the turbulent kinetic energy, κ, and the rate of dissipation ε, respectively, they are given by:

$$U_i \frac{\partial k}{\partial x_i} = \frac{\partial}{\partial x_i}\left(\frac{\mu_t}{\rho \sigma_k}\frac{\partial k}{\partial x_i}\right) + P_k - \varepsilon \tag{4}$$

$$U_i \frac{\partial \varepsilon}{\partial x_i} = \frac{\partial}{\partial x_i}\left(\frac{\mu_t}{\rho \sigma_\varepsilon}\frac{\partial \varepsilon}{\partial x_i}\right) + c_1 \frac{\varepsilon}{k}P_k - c_2 f_2 \frac{\varepsilon^2}{k} \tag{5}$$

The symbols P_k and μ_t, represent the rate of the turbulent kinetic energy production and the turbulent viscosity, respectively, are expressed by:

$$P_k = \tau_{ij}\frac{\partial U_i}{\partial x_j}, \; \mu_t = c_\mu f_\mu \rho \frac{k^2}{\varepsilon} \tag{6}$$

In the present work for NLEVM, the formulations of Low Reynolds Number will be assumed for wall treatment. The damping functions f_2 and f_μ observed in Equations (5) and (6) were proposed by Abe et al [36]. These functions and the constant c_1 and c_2 are used in equations $k - \varepsilon$. The subscript P refers to the nodal point near to the wall. Thus U_P and k_P are the values of the velocity and kinetic energy in this point, respectively. For the constants c_μ, c_1, c_2, σ_k and σ_ε are assumed the values of 0.09; 1.5; 1.9; 1.4 e 1.3; respectively. New constitutive relation for the tensions of Reynolds in the NLEVM model was assumed in according to the thesis of Assato [8]:

$$\tau_{ij} = \left(\mu_t S_{ij}\right)^L + \left(c_{1NL}\,\mu_t \frac{k}{\varepsilon}\left[S_{ik}S_{kj} - \frac{1}{3}S_{kl}S_{kl}\delta_{ij}\right]\right)^{NL} \tag{7}$$

This expression shows that the second term of the right side in Equation (7) represents the nonlinear relation added the original constitutive relation. This quadratic term represents the degree of anisotropy between the normal tensions of Reynolds responsible for predicting the secondary flow in non circular ducts. The values of c_{1NL} proposed by Speziale [35] is equal to 1.68. In this work, c_{1NL} will be analyzed and will adopt values for the formulation of Low Reynolds Numbers. The normal and shear tensions of Reynolds are expressed as:

$$\tau_{xx} = c_{1NL}\mu_t\frac{k}{\varepsilon}\left[\frac{1}{3}\left(\frac{\partial W}{\partial x}\right)^2 - \frac{2}{3}\left(\frac{\partial W}{\partial y}\right)^2\right]; \; \tau_{yy} = c_{1NL}\mu_t\frac{k}{\varepsilon}\left[\frac{1}{3}\left(\frac{\partial W}{\partial y}\right)^2 - \frac{2}{3}\left(\frac{\partial W}{\partial x}\right)^2\right] \tag{8}$$

$$\tau_{xy} = c_{1NL}\mu_t\frac{k}{\varepsilon}\left[\frac{\partial W}{\partial x}\frac{\partial W}{\partial y}\right]; \; \tau_{xz} = \mu_t\frac{\partial W}{\partial x}; \; \tau_{yz} = \mu_t\frac{\partial W}{\partial y} \tag{9}$$

The following differences for the normal tensions of Reynolds are presented and used in order to predict the anisotropy in turbulent flow at non circular ducts,

$$\left(\tau_{yy} - \tau_{xx}\right) = c_{1NL}\mu_t \frac{k}{\varepsilon}\left[\left(\frac{\partial W}{\partial y}\right)^2 - \left(\frac{\partial W}{\partial x}\right)^2\right] \tag{10}$$

Therefore, the Equation (6), including the tensions of Reynolds given in Equation (9), the turbulence production term is expressed as:

$$P_k = \tau_{xz}\frac{\partial W}{\partial x} + \tau_{yz}\frac{\partial W}{\partial y} \tag{11}$$

2.2.2. Reynolds Stress Model (RSM)

The most complex turbulence model is the Reynolds Stress Model (RSM), also known as second order model. It involves calculations of Reynolds stresses to an individual form, $\overline{\rho u_i' u_j'}$. These Reynolds stresses are used for formulating the differential equations of turbulent flow transport. The individual Reynolds stresses are utilized to close the average Reynolds equations of the momentum conservation. This model has shown superiority in relation to the two equation models (for example, $k - \varepsilon$) in simulating of complex flows that involve swirl, rotation, etc. The exact transport equations for the Reynolds stresses, $\overline{\rho u_i' u_j'}$, can be written as:

$$\frac{\partial}{\partial t}(\overline{\rho u_i' u_j'})(a) + \frac{\partial}{\partial x_k}(\rho u_k \overline{u_i' u_j'})(b) = -\frac{\partial}{\partial x_k}\left[\overline{\rho u_i' u_j' u_k'} + \overline{p(\delta_{kj}u_i' + \delta_{ik}u_j')}\right](c) +$$

$$\frac{\partial}{\partial x_k}\left[\mu\frac{\partial}{\partial x_k}(\overline{u_i' u_j'})\right](d) - \rho\left(\overline{u_i' u_k'}\frac{\partial u_j}{\partial x_k} + \overline{u_j' u_k'}\frac{\partial u_i}{\partial x_k}\right)(e) - \rho\beta\left(g_i\overline{u_j'\theta} + g_j\overline{u_i'\theta}\right)(f) + \tag{12}$$

$$\overline{p\left(\frac{\partial u_i'}{\partial x_j} + \frac{\partial u_j'}{\partial x_i}\right)}(g) - 2\mu\overline{\frac{\partial u_i'}{\partial x_k}\frac{\partial u_j'}{\partial x_k}}(h) - 2\rho\Omega_k\left(\overline{u_j' u_m'}\varepsilon_{ikm} + \overline{u_i' u_m'}\varepsilon_{jkm}\right)(i) + S(j)$$

Where the letters represent: (a) Local derivative of the time; (b) $C_{ij} \equiv$ Convection; (c) $D_{T,ij} \equiv$ Turbulent diffusion; (d) $D_{L,ij} \equiv$ Molecular diffusion; (e) $P_{ij} \equiv$ Production term of stresses; (f) $G_{ij} \equiv$ Buoyancy production term; (g) $\phi_{ij} \equiv$ Pressure-stress (redistribution); (h) $\varepsilon_{ij} \equiv$ Dissipation term; (i) $F_{ij} \equiv$ Production term for the rotation system; (j) $S_j \equiv$ Source term. The terms of the exact equations presented previously, C_{ij}, $D_{L,ij}$, P_{ij} and F_{ij} do not require modeling. However, the terms $D_{T,ij}$, G_{ij}, ϕ_{ij} and ε_{ij} need to be modeled to close the equations. For the present analysis, the model LRR (Launder, et al [37]) is chosen, which assumes that the correlation of velocity- pressure is a linear function of anisotropy tensor in the phenomenology of the redistribution, ϕ_{ij}. For the wall treatment, it is also assumed the Low Reynolds numbers and periodic conditions in according to Rokni [16]. This model had been simulated in the commercial code Fluent 6.3.

2.3 Turbulence models for turbulent heat flux

2.3.1. Simple Eddy Diffusivity (SED)

This method is based on the Boussinesq viscosity model. The turbulent diffusivity for the energy equation can be expressed as: $\alpha_t = \dfrac{\mu_t}{\rho \sigma_t}$, where the turbulent Prandtl number, σ_t for the SED model assumes a value constant in the entire region. For the air, σ_t it assumes a value equal to 0.89. The turbulent heat flux is given by,

$$\overline{\rho u_j t} = -\frac{\mu_t}{\sigma_T}\frac{\partial T_f}{\partial x_j} \tag{13}$$

2.3.2. Generalized Gradient Diffusion Hypothesis (GGDH)

Daly and Harlow [38] present the following formulation to the turbulent heat flux:

$$\overline{\rho u_j t} = -\rho C_t \frac{k}{\varepsilon}\left(\overline{u_j u_k}\frac{\partial T_f}{\partial x_k}\right) \tag{14}$$

The constant C_t, assumes the value of 0.3. The main advantage of this model is in considering the anisotropic behavior of the fluid heat transport in ducts.

2.3.3. Dimensionless energy equation for SED and GGDH models

For a given cross section of area "A", it is possible to define a mean velocity "U_b" and a bulk temperature "T_b", expressed as:

$$U_b = \frac{1}{A}\iint W.dx.dy \tag{15}$$

and

$$T_b = \frac{\displaystyle\int_A W.T_f.dA}{U_b.A} = \frac{1}{A.U_b}\iint W.T_f.dx.dy \tag{16}$$

Kays and Crawford [30] developed a formulation to rectangular cross section ducts. They considered the boundary conditions with prescribed uniform wall temperatures at the cross section, and at the duct length. According to Garcia [31], it is possible to carry out an analysis with non-uniform wall temperature boundary conditions. In this case, it is necessary to define a value that represents the mean wall temperatures in a given cross section, "T_{Wm}", given as:

$$T_{Wm} = \frac{\left[\frac{1}{L}\int_0^L T_1(0,y).dy + \frac{1}{D}\int_0^D T_2(x,0).dx + \frac{1}{L}\int_0^L T_3(D,y).dy + \frac{1}{D}\int_0^D T_4(x,L).dx\right]}{2(L+D)} \qquad (17)$$

It is possible to develop a formula similar to Kays and Crawford [30], and a new expression for the turbulent energy equation can be presented as:

$$U\frac{\partial T_f}{\partial x} + V\frac{\partial T_f}{\partial y} + W\frac{\partial T_f}{\partial z} - \left[\frac{\partial}{\partial x}\left(\alpha\frac{\partial T_f}{\partial x} - \overline{ut}\right) + \frac{\partial}{\partial y}\left(\alpha\frac{\partial T_f}{\partial y} - \overline{vt}\right) + \frac{\partial}{\partial z}\left(\alpha\frac{\partial T_f}{\partial z} - \overline{wt}\right)\right] = 0 \qquad (18)$$

The following considerations are applied to obtain the variables in dimensionless form:

$$X = \frac{x}{D_h}, \qquad (19)$$

$$Y = \frac{y}{D_h} \qquad (20)$$

and

$$\phi = \frac{\alpha.\left(T_{Wm} - T_f\right)}{U_b.D_h^2.\left(\frac{dT_b}{dz}\right)} \qquad (21)$$

Replacing the Equations (13) or (14), (19)-(21) in Equation (18), dimensionless energy equations for SED and GGDH are obtained, respectively, expressed as:

$$\frac{\partial}{\partial X}\left\{(\alpha + \alpha_t)\frac{\partial\phi}{\partial X}\right\} + \frac{\partial}{\partial Y}\left\{(\alpha + \alpha_t)\frac{\partial\phi}{\partial Y}\right\} - (D_h)\left(U\frac{\partial\phi}{\partial X} + V\frac{\partial\phi}{\partial Y}\right) = -\frac{W}{U_B}\alpha\left(\frac{\phi}{\phi_B}\right) \qquad (22)$$

$$\frac{\partial}{\partial X}\left[(\alpha_{ex})\frac{\partial\phi}{\partial X}\right] + \frac{\partial}{\partial Y}\left[(\alpha_{ey})\frac{\partial\phi}{\partial Y}\right] - D_h\left(U\frac{\partial\phi}{\partial X} + V\frac{\partial\phi}{\partial Y}\right) = -\frac{W}{U_B}\alpha\left(\frac{\phi}{\phi_B}\right) - C_t\frac{\partial}{\partial X}\left[\Gamma_X\frac{\partial\phi}{\partial Y}\right]$$
$$-C_t\frac{\partial}{\partial Y}\left[\Gamma_Y\frac{\partial\phi}{\partial X}\right] \qquad (23)$$

The fluid temperature field "T_f" can be replaced by "T_b" and the Equation (21) can be expressed as:

$$\phi_b = \frac{\alpha.\left(T_{Wm} - T_b\right)}{U_b.D_h^2.\left(\frac{dT_b}{dz}\right)}, \qquad (24)$$

and

$$T_b = T_{Wm} - \frac{D_h^2.U_b}{\alpha} \cdot \frac{dT_b}{dz} \cdot \phi_b \qquad (25)$$

From Equation (21), the Equation (26) is obtained, and applying this in Equation (16), the bulk temperature is obtained and expressed in Equation (27):

$$T_f = T_{Wm} - \frac{\phi.U_b.D_h^2.\left(\dfrac{dT_b}{dz}\right)}{\alpha}, \qquad (26)$$

and

$$T_b = T_{Wm} - \frac{D_h^2}{\alpha.A} \cdot \frac{dT_b}{dz} \cdot \iint W.\phi.dx.dy \qquad (27)$$

Replacing Equation (27) in Equation (24), and using Equations (19) and (20), the dimensionless bulk temperature is given as:

$$\phi_b = \frac{D_h^2}{A.U_b} \cdot \iint W.\phi.dX.dY \qquad (28)$$

It is possible to compute the heat transfer rate per unit length on the wall surface, " q' ", as shown in Equation (29) in function of " T_{Wm} ", " T_b ", and the average heat convection coefficient, " $\bar{h}$ ". From fluid enthalpy derivative gradient [$dh_b = c_p.dT_b$], the heat transfer rate per unit length in the fluid, " q'_f ", can be expressed by Equation (30).

$$q' = P_e.\bar{h}.(T_{Wm} - T_b), \qquad (29)$$

and

$$q'_f = \rho.U_b.A.c_p.\frac{dT_b}{dz} \qquad (30)$$

Equation (30) can be integrated to two cross sections (inlet, z_1, and outlet, z_2), thus, the following expression is obtained,

$$T_{b_{z2}} = T_{Wm} - \left(T_{Wm} - T_{b_{z1}}\right).e^{-\left(\frac{P_e}{2A}\right)^2 \frac{\alpha}{U_b}.Nu.(z_1 - z_2)} \qquad (31)$$

From Equation (31), the bulk temperature longitudinal (z-$axis$) variation "ΔT_b" is obtained. It is done by "cutting" the duct into a lot of segments and applying the numerical method to find "ΔT_b" at each finite cross section. For a given bulk temperature at the duct inlet section (T_{b1}), after solving the equation system, duct outlet bulk temperature (T_{b2}) is calculated from Equation (31). The Dimensionless boundary conditions are given by the following equations:

$$\phi(0,Y) = \frac{\alpha.\left(T_{Wm} - T_1\right)}{U_b.D_h^2.\left(\dfrac{dT_b}{dz}\right)} \tag{32a}$$

and

$$\phi(X,0) = \frac{\alpha.\left(T_{Wm} - T_2\right)}{U_b.D_h^2.\left(\dfrac{dT_b}{dz}\right)} \tag{32b}$$

$$\phi(\tfrac{D}{D_h},Y) = \frac{\alpha.\left(T_{Wm} - T_3\right)}{U_b.D_h^2.\left(\dfrac{dT_b}{dz}\right)}, \text{ and } \phi(X,\tfrac{L}{D_h}) = \frac{\alpha.\left(T_{Wm} - T_4\right)}{U_b.D_h^2.\left(\dfrac{dT_b}{dz}\right)} \tag{33}$$

When considering uniform wall temperature, the Equations (32) and (33) are equal to zero, and for these particular conditions, it is possible to notice that these boundary conditions are not functions of " dT_b/dz ". That simplification becomes equal to the one studied by Patankar [32]. Equations (24)-(26) , as well as the boundary conditions from Equations (32) and (33), form a set of differential equations, in which "ϕ" and "dTb/dz" parameters are unknown. When that equation system is solved, it is possible to obtain "T_f".

2.3.4. Additional equations

Additional equations were utilized for the calculation of the factor of friction Moody, f ; coefficient of friction Fanning, C_f ; Prandtl law; local Nusselt number for the Low Reynolds formulation (Rokni [16]), Nu_{xp} and Correlation of Gnielinsky, respectively. Thus, the additional equations are given by the following equations:

$$f \equiv \frac{-\left(dP/dz\right).D_h}{\rho U_B^2 \Big/ 2}, \tag{34}$$

$$C_f = \frac{f}{4} \tag{35}$$

$$\frac{1}{\sqrt{f}} = 2\log\left(\text{Re}\sqrt{f}\right) - 0.8, \tag{36}$$

$$Nu_{xp} = D_h \frac{\left(T_w - T_p\right)}{\eta\left(T_w - T_b\right)} \tag{37}$$

$$Nu = \left[\frac{(f/8)(\text{Re}-1000)\text{Pr}}{\left[1+12.7(f/8)^{1/2}\left(\text{Pr}^{2/3}-1\right)\right]} \right]$$

(38)

3. Numerical implementation

After applying the method of finite differences to the algebraic equations, to obtain the temperature fields, the following five steps indicate the developed methodology in the numerical solution. (Garcia [31]):

Step 1. To define the function value of the non uniform temperatures in the walls of the duct $T_{Wm} = f\left(T_1(0,y), T_2(x,0), T_3(D,y), T_4(x,L)\right)$, This function can be expressed by a Fourier expansion;

Step 2. To obtain velocity field and estimated values for "U_b" "T_{Wm}" and "dT_b/dz";

Step 3. Equations for the boundary conditions are evaluated (Equations 32 and 33);

Step 4. Dimensionless energy equation (Temperature field, "$\emptyset$") the Equation (23) is solved and "$\emptyset_b$" is computed according to Equation (28), until convergence is obtained ($\emptyset_b$ < tolerance). This is the end of the first iterative loop;

Step 5. A value for "dT_b/dz" is computed in accordance with Equation (25). Boundary conditions are updated (step 3) to obtain a solution for the new temperature field (step 4), until convergence is obtained (dT_b/dz < tolerance). This is the end of the second iterative loop;

For all steps, "tolerance of 10^{-7}" is the value to be accomplished by the convergence criteria, which is applicable to "$\emptyset_b$" (dimensionless bulk temperature), "dT_b /dz" and "$\emptyset$" (dimensionless temperature field).

4. Results and discussion

4.1. Fluid flow and heat transfer field

The Figure 2(a) shows the utilized grid (120X120) in the numerical simulation for the formulations of Low Reynolds, the Figure 2(b) it represents the secondary flow contours and comparisons of the velocity profile (NLEVM, Assato [8]) with the experimental work of the Melling and Whitelaw [5] for fluid water and Re=42000.

The predicted distributions of the friction coefficient (NLEVM and RSM) and Nusselt number (SED and GGDH) dependence on Reynolds number for fully developed flow and heat transfer in a square duct is shown in Figure 3(a) and 3(b), respectively.

Figure 4(a): comparisons of the Results (RSM-SED) numerical with the experimental for temperature profile (wall constant temperature) $(T_{Wm} - T_f)/(T_{Wm} - T_C)$ with fluid air and Re=65000 (Hirota [15]) are shown, the figure 4(b) shows the variation of the temperature profile with non-uniform wall Temperature: south=400K, north=373K, east=393K, west=353K; presented as Case I.

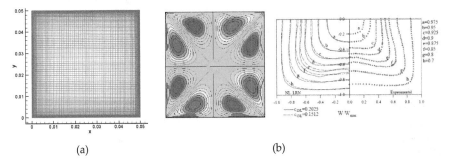

(a)   (b)

Figure 2. (a) Grid 120X120 for numerical simulation (b) Secondary flow contours and comparisons of the axial mean velocity with Melling and Whitelaw [5] for water utilizing NLEVM Model.

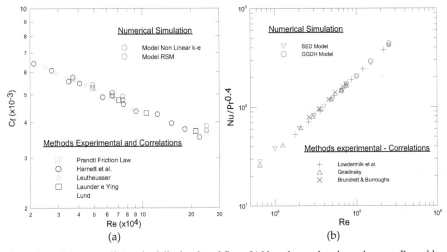

(a)   (b)

Figure 3. (a) Friction coefficient for fully developed flow, (b) Nusselt number dependence on Reynolds number for fully developed flow

Already the Figure 5(a) shows: The variation of the temperature profile with non-uniform wall temperature, represented by means of functions sine (Case II), south=$(350-20Sin(\zeta))$K, north=$(400-50Sin(\zeta))$K, east=$(330+20Sin(\zeta))$K, west=$(350+50Sin(\zeta))$K. where ζ is function of the radians $(0-\pi/2)$ and i,j (points number of the grid in the direction x and y, respectively). The Figure 5(b) shows the behavior of the "T_b" and "DTb/dz" for different square cross sections in the direction of the main flow, according to Equation (31).

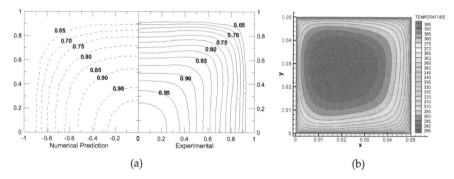

(a)   (b)

Figure 4. (a) Results (RSM-SED) numerical and (Hirota et al [15]) experimental for mean temperature (uniform wall temperature) (b) Fluid temperature with non-uniform wall Temperature (Case I).

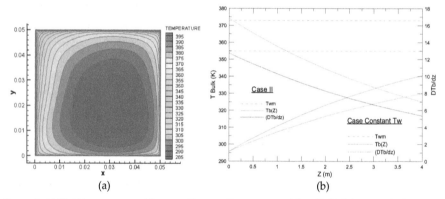

(a)   (b)

Figure 5. (a) Fluid temperature with non-uniform wall temperature (Case II) (b) Behavior "T_b" for different square cross sections and cases in the direction of the main flow.

The Figures 6 (a) and (b), shown the temperature distribution for a rectangular duct aspect ratio (1:2) represented by means of a function sine (Case II). A third case denominated Case III is represented by: south= $405 + 10\left[\dfrac{nx\text{-}1}{nx_{máx}-1}\right]$ K; North= $395 + 10\left[\dfrac{nx\text{-}1}{nx_{máx}-1}\right]$ K; east= $405\text{-}10\left[\dfrac{ny\text{-}1}{ny_{máx}-1}\right]$ K; west= $415 - 10\left[\dfrac{ny\text{-}1}{ny_{máx}-1}\right]$ K. Some results for rectangular ducts are shown in the Table 1.

In the doctoral thesis Garcia (1996) was analyzed the laminar flow coupled to the conduction and radiation in rectangular ducts and concluded that as increases the aspect ratio, the Nusselt number found in the coupling, differs from that found for ducts with constant temperature imposed around the perimeter of the section. which shows that could be making a mistake to consider the literature results without calculating the energy equation.

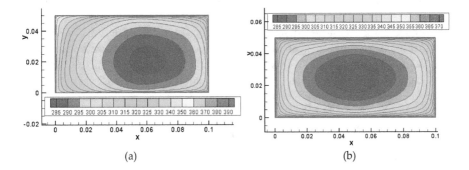

(a)   (b)

Figure 6. (a) Retangular duct aspect ratio (1:2) case II, Re=65000 with Tb=300 K, utilizing the SED model (b) Retangular duct aspect ratio (1:2) with constant temperature in the perimeter Twm=373 K, Re=65000 with Tb=300 K, utilizing the SED model.

In the present study, the variations of the average Nusselt number for a square duct and different cases analyzed (uniform and non uniform temperature in the perimeter) are minimal. Already in the case of rectangular duct with an aspect ratio (1:2), the variations should be taken into account as shown in Table 1.

Cases Analyzed	Reynolds number (Re)	Nusselt number calculated (Nu)	Correlation Dittus Boelter (Nu).
Temperature Constant	65000	145, 910	142,89
Case II	65000	139, 682	-
Case III	65000	145, 059	-
Temperature Constant	28853	79, 101	77,1
Case III	28853	76, 769	-

Table 1. Numerical results obtained through RSM-SED model [34], for the averaged Nusselt number in a rectangular duct with aspect ratio (1:2).

5. Conclusions

The results shown what for the friction factor and Nusselt number in a wide range of the Reynolds number with uniform wall temperature have a reasonable approach with the experimental works and correlation of the literature, (Figure 3(a), (b)). The Figures 4(b) and 5(a) shown new results investigated in present study, which is observed a distortion of the temperatures field and as consequence a variation of the Nusselt number caused mainly by the distribution of the non-uniform wall temperature (Case I and II, with fluid air and $Re=65000$, respectively). Most applications can be approximated by the functions sine and cosine in the wall, but we are able to resolve by means of the methodology presented, any peripheral heat flux variation that can be expressed by a Fourier expansion (Kays and Crawford [30]). The Figure 5(b) shows the comparisons of the behavior of the curves "T_b" and "DT_b/dz" in the direction of the main flow for Case II and Case uniform wall temperature. The variations of the average Nusselt number for a square duct and different cases analyzed (uniform and non uniform temperature in the perimeter) are minimal. Already in the case of duct with an aspect ratio (1:2) the variations should be taken into account. These results can be helpful in the project of thermal devices as in heat transfer and secondary flows in cavities, seals, channel of gas turbines and others.

Nomenclature

C_{ij}	convection
c_p	specific heat at constant pressure
D	duct height
D_h	hydraulic diameter, $D_h = 4.A/P_e = 2.L.D/(L+D)$
$D_{L,ij}$	molecular diffusion
$D_{T,ij}$	Turbulent diffusion
dP/dz	pressure gradient at z direction (longitudinal axis)
f, C_f	factor of Moody's friction, and Fanning's friction coefficient, respectively.
F_{ij}	Term production for the rotation system
G_{ij}	buoyancy production Term
k_f	fluid thermal conductivity
L	duct width
Nu	Nusselt number
P_e	perimeter
P_{ij}	production Term of tensions
P_k	turbulence production term
Re	Reynolds number
S_j	Source term
T	temperature
T_b	internal flow bulk temperature
T_{Wm}	wall mean temperature
T_1, T_2, T_3 and T_4	temperature distributions at duct wall (bottom, right side, top and left side)
U_b	internal flow bulk velocity
U, V and W	average velocity in the direction x, y and z; respectively
y^+	dimensionless wall distance

Greek Symbols

α	thermal diffusivity, $\alpha = k_f / (\rho. c_p)$
ε_{ij}	Term of dissipation
η	distance normal to the wall
ϕ	dimensionless temperature distribution
ϕ_{ij}	Term of pressure-tension (redistribution)
μ	dynamic viscosity
μ_t	turbulent viscosity
ρ	density
σ_t	turbulent Prandtl number

Author details

G.A. Rivas[*], E.C. Garcia and M. Assato
Instituto Tecnologico de Aeronautica (ITA),Brazil

6. References

[1] Z.H. Qin; R.H. Plecther, Large eddy simulation of turbulent heat transfer in a rotating square duct, International Journal Heat Fluid Flow. 27 (2006) 371-390.

[2] J. Nikuradse, 1926, Untersuchung uber die Geschwindigkeitsverteilung in turbulenten Stromungen, Diss. Göttingen, VDI - forschungsheft 281.

[3] F.B. Gessner and A.F. Emery, A Reynolds stress model for turbulent corner flows – Part I: Development of the model, Journal Fluids Eng. 98 (1976) 261-268.

[4] F.B. Gessner, and J.K. Po, A Reynolds stress model for turbulent corner flows – Part II: Comparison between theory and experiment, Journal Fluids Eng. 98 (1976) 269-277.

[5] A. Melling and J.H. Whitelaw, Turbulent flow in a rectangular duct, Journal Fluid Mechanical. 78 (1976) 289-315.

[6] A. Nakayama, A.; W.L. Chow and D. Sharma, Calculation of fully development turbulent flows in ducts of arbitrary cross-section, Journal Fluid Mechanical, 128 (1983) 199-217.

[7] H.K. Myon and T. Kobayashi, Numerical Simulation Of Three Dimensional Developing Turbulent Flow in a Square Duct with the Anisotropic κ-ε Model, Advances in Numerical Simulation of Turbulent Flows ASME, Fluids Engineering Conference, 1991. Vol.117, Portland, United States of America, pp. 17-23.

[8] M. Assato, Análise numérica do escoamento turbulento em geometrias complexas usando uma formulação implícita, Doctoral Thesis, Departamento de Engenharia

---

* Corresponding Author

Mecânica, Instituto Tecnológico de Aeronáutica - ITA, São José dos campos - SP, Brazil, 2001.

[9] M. Assato; M.J.S. De Lemos, Turbulent flow in wavy channels simulated with nonlinear models and a new implicit formulation, Numerical Heat Transfer – Part A: Applications. 56 (4) (2009) 301-324.

[10] D. Home; M.F. Lightstone; M.S. Hamed, Validation of DES-SST based turbulence model for a fully developed turbulent channel flow problem, Numerical Heat Transfer – Part A: Applications. 55 (4) (2009) 337-361.

[11] D. D. Luo; C.W. Leung; T.L. Chan; W.O. Wong, Simulation of turbulent flow and forced convection in a triangular duct with internal ribbed surfaces, Numerical Heat Transfer – Part A: Applications. 48 (5) (2005) 447-459.

[12] S. Ergin; M. Ota; H. Yamaguchi, Numerical study of periodic turbulent flow through a corrugated duct, Numerical Heat Transfer – Part A: Applications. 40 (2) (2001) 139-156.

[13] B.E. Launder and W.M.Ying., Prediction of flow and heat transfer in ducts of square cross section", Proc. Inst. Mech. Eng., 187 (1973) 455-461.

[14] A.F. Emery, P.K. Neighbors and F.B. Gessner, The numerical prediction of developing turbulent flow and heat transfer in a square duct, Journal Heat Transfer, 102 (1980) 51–57.

[15] M. Hirota, H. Fujita, H. Yokosawa, H. Nakai, H. Itoh, Turbulent heat transfer in a square duct, International Journal Heat and fluid flow, 18 (1997) 170-180.

[16] M. Rokni, Numerical investigation of turbulent fluid flow and heat transfer in complex duct, Doctoral Thesis, Department of Heat and Power Engineering. Lund Institute of Technology, Sweden, 1998.

[17] Y. Hongxing, Numerical study of forced turbulent heat convection in a straight square duct, International Journal of Heat and Mass Transfer, 52 (2009) 3128-3136.

[18] Y.T. Yang; M.L. Hwang, Numerical simulation of turbulent fluid flow and heat transfer characteristics in a rectangular porous channel with periodically spaced heated blocks, Numerical Heat Transfer – Part A: Applications. 54 (8) (2008) 819-836.

[19] T.S. Park, Numerical study of turbulent flow and heat transfer in a convex channel of a calorimetric rocket chamber, Numerical Heat Transfer – Part A: Applications. 45 (10) (2004) 1029-1047.

[20] J. Zhang; L. Dong; L. Zhou; S. Nieh, Simulation of swirling turbulent flows and heat transfer in a annular duct, Numerical Heat Transfer – Part A: Applications. 44 (6) (2003) 591-609.

[21] B. Zheng; C.X. Lin; M.A. Ebadian, Combined turbulent forced convection and thermal radiation in a curved pipe with uniform wall temperature, Numerical Heat Transfer – Part A: Applications. 44 (2) (2003) 149-167.

[22] J. Su; A.J. Da Silva Neto, Simultaneous estimation of inlet temperature and wall heat flux in turbulent circular pipe flow, Numerical Heat Transfer – Part A: Applications. 40 (7) (2001) 751-766.

[23] A. Saidi; B. Sundén, Numerical simulation of turbulent convective heat transfer in square ribbed ducts, Numerical Heat Transfer – Part A: Applications. 38 (1) (2001) 67-88.

[24] M. Rokni, A new low-Reynolds version of an explicit algebraic stress model for turbulent convective heat transfer in ducts, Numerical Heat Transfer – Part B: Fundamentals. 37 (3) (2000) 331-363.

[25] A. Valencia, Turbulent flow and heat transfer in a channel with a square bar detached from the wall, Numerical Heat Transfer – Part A: Applications. 37 (3) (2000) 289-306.

[26] M.C. Sharatchandra; D.L. Rhode, Turbulent flow and heat transfer in staggered tube banks with displaced tube rows, Numerical Heat Transfer – Part A: Applications. 31 (6) (1997) 611-627.

[27] A. Campo; K. Tebeest; U. Lacoa; J.C. Morales, Application of a finite volume based method of lines to turbulent forced convection in circular tubes, Numerical Heat Transfer – Part A: Applications. 30 (5) (1996) 503-517.

[28] M. Rokni; B. Sundén, Numerical investigation of turbulent forced convection in ducts with rectangular and trapezoidal cross section area by using different turbulence models, Numerical Heat Transfer – Part A: Applications. 30 (4) (1996) 321-346.

[29] G. Yang; M.A. Ebadian, Effect of Reynolds and Prandtl numbers on turbulent convective heat transfer in a three-dimensional square duct, Numerical Heat Transfer – Part A: Applications. 20 (1) (1991) 111-122.

[30] W.M. Kays; M.Crawford, Convective Heat and Mass Transfer, McGraw-Hill, New York, USA, 1980, pp. 250-252.

[31] E.C. Garcia, Condução, convecção e radiação acopladas em coletores e radiadores solares, Doctor degree thesis, ITA - Instituto Tecnológico de Aeronáutica, São José dos Campos, SP, Brasil, 1996.

[32] S.V. Patankar, Computation of Conduction and Duct Flow Heat Transfer, Innovative Research, Maple Grove, USA, 1991.

[33] M.J. Moran, et al, Introdução à Engenharia de Sistemas Térmicos: Termodinâmica, Mecânica dos Fluidos e Transferência de Calor, LTC Ed., Rio de Janeiro-RJ, Brasil, 2005.

[34] G. A. Rivas Ronceros, Simulação numérica da convecção forçada turbulenta acoplada à condução de calor em dutos retangulares, Doctor degree thesis, ITA - Instituto Tecnológico de Aeronáutica, São José dos Campos, SP, Brasil, 2010.

[35] C. G. Speziale, On Nonlinear k-ε and k-l Models of Turbulence, J. Fluid Mech., vol. 176, pp. 459-475, 1987.

[36] K. Abe, et al, An Improved k-ε Model for Prediction of Turbulent Flows with Separation and Reattachment, Trans. JSME, Ser. B, vol. 58, pp. 3003-3010, 1992.

[37] B.E. Launder, G.J. Reece, and W. Rodi: Progress in the development of a Reynolds-stress turbulence closure, J. Fluid Mech., vol.68, 537-566, 1975.

[38] Daly and Harlow: transport equations in turbulence, Phys. Fluids., vol.13, 2634-2649, 1970.

Boiling and Condensation

Chapter 8

Critical Heat Flux in Subcooled Flow Boiling of Water

Yuzhou Chen

Additional information is available at the end of the chapter

1. Introduction

Flow boiling has an extremely high heat transfer coefficient, and is applied in variety of practices. However, once the heat flux exceeds a certain high level the heated surface can no longer support continuous liquid contact, associated with substantial reduction in the heat transfer efficiency. It may result in a sudden rise of surface temperature in a heat flux controlled system, or a drastic decrease in power transferred in a temperature controlled system. This phenomenon is called the boiling crisis, and the maximum heat flux just before the boiling crisis is usually referred to as critical heat flux (CHF).

Depending on the flow regimes, two types of CHF are classified: (i) in subcooled or low quality region the CHF is characterized by the transition from nucleate boiling to film boiling, and it is termed as the departure from nucleate boiling (DNB); (ii) in higher quality region the CHF is characterized by the dryout of liquid film of annular flow. The DNB and dryout have substantially different mechanisms, and are generally cataloged as the first and the second kind of critical heat flux, respectively (Tong and Tang, 1997).

The CHF is an important subject to many engineering applications. Especially, in a nuclear reactor the occurrence of critical heat flux could lead to a failure of fuel element, and thus the CHF is a major limit for the reactor safety. During past five decades the CHF has been investigated extensively over the world theoretically and experimentally (IAEA-TECDOC-1203, 2001). A great number of empirical correlations and physical models have been proposed. In recent years, a Look-Up Table method (LUT) is widely accepted due to its advantages of higher accuracy, wider range of application, correct asymptotic trend and convenience for updating (Groeneveld, et al., 1996). This table is applied in the system code RELAP5 for reactor safety analysis. The LUT contains a tabulation of normalized CHF data of a uniformly heated tube of 8 mm in diameter at discrete local parameters of pressure, mass flux and quality. Several correction factors are incorporated for utilization of the LUT in other conditions to account for the effects of diameter, bundle, spacer, flux distribution, flow orientation, etc.. Unfortunately, there exists a scarcity of CHF data in low pressure/low flow/ subcooled region, as shown in Fig.1. Because of the extreme complexity of the phenomenon and the lack of adequate knowledge of the mechanisms, all these predictive methods are heavily relied on experimental data, and can not be extrapolated out of their ranges with confidence.

In China Institute of Atomic Energy (CIAE), in the past four decades a great number of CHF experimental data of subcooled boiling of flowing water were obtained in tubes and annuli at lower pressure with different diameter or gap to support the designs of research reactors, HFR and CARR, which were first put into operation in 1980 and 2011, respectively. In recent years the experiments were extended to the region of near-critical pressure with lower flow for the R&D of Supercritical Water-Cooled Reactor (SCWR). In these experiments the CHF (DNB) phenomena are studied with emphases on lower pressure and higher pressure with lower flow. The characteristics and parametric trends of the CHF are clarified, and the physical models are derived.

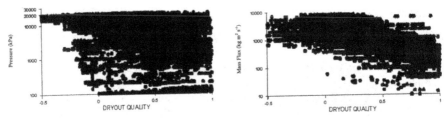

Figure 1. Range of database for 1996 CHF Look-up table (Groeneveld, et al., 1996)

2. Parametric trends

A comprehensive review on the subcooled flow boiling CHF was given by Berglest (1977). As concluded, the effect of heating length disappears when exceeding 200 mm and the subcooled boiling CHF depends only on local parameter ("local condition hypothesis"). The CHF increases with the local mass flux and subcooling increasing. The effect of pressure is more complicated: the CHF increases with pressure in low pressure region, is relatively constant over an intermediate range of pressure, and decreases in high pressure region. The effect of geometry on CHF is observed in many experiments, and is accounted by a factor $f = (D_e/8)^n$, in which D_e is the hydraulic diameter in mm. Different value of the exponent n is obtained from different experiments, ranging from -0.1 to -0.5. For applying the LUT in reactor safety analysis code RELAP5 the value of n = -0.5 is used. While for annulus or rectangular channels the effect of gap width is not observed in many experiments.

In Savannah River lab and Columbia University the subcooled CHF experiments were conducted with D_2O and H_2O coolant and aluminum and stainless-steel heaters (Knoebel et al., 1973). It was concluded that the CHF for D_2O is 16% greater than for H_2O at constant subcooling and velocity, and the CHF of aluminum heaters is a minimum of 20% greater than for stainless steel.

Some typical subcooled boiling CHF correlations are listed in Tab.1. Similar trends of the CHF with the flow rate and subcooling are represented in various correlations, but the degrees of these effects are different significantly. The present investigation has revealed that these effects are inter-dependent, associated with complicated parametrical trends for different regions of conditions.

Author	Correlation	Range of parameters
Knoebel et al. (1973)	$q_c = 0.485(1 + 0.169V)(1 + 0.124\Delta T_s)$ for H_2O and Stainless-steel annuli and rectangular channels	D_e=5.1-9.5mm,L/D=64, p=0.2-0.66MPa,ΔT_s =25 - 90K, V=3.9-18.3m/s
Mirshak et al. (1959)	$q_c = 1.51(1 + 0.12V)(1 + 0.00914\Delta T_s)(1 + 1.9p)$	D_e=5.3- 11.7mm,p=0.17- 0.6 MPa,ΔT_s = 5-75K, V=1.5-13.7m/s
Zeigarnik (1994)	$q_c = A + B(G/1000)\Delta T_s$ A and B are given in curves in relation with D_e	D_e=3-14mm, L/D=50- 62.5, p=0.5-3.0MPa,ΔT_s =20- 160K, G=4- 25Mg/m²s
Sudo and Kaminaga (1993)	$q^*_{c,1} = 0.005\left|G^*\right|^{0.611}(1 + 5000\Delta T^*_0 /\left|G^*\right|)$ for $G \geq G^*_1$ $q^*_{c,2} = (A/A_H)(\Delta h_{in}/h_{fg})G^*$ for $G^*_2 \leq G < G^*_1$ $q^*_{c,3} = 0.7(A/A_H)(W/\lambda)^{0.5}/(1 + (\rho_g/\rho_l)^{0.25})^2$ for $G < G^*_2$ $q^*_c = q_c/[h_{fg}(\lambda g \rho_g(\rho_l - \rho_g))^{0.5}], \lambda^* = [\sigma/(g(\rho_l - \rho_g))]^{0.5}$ $G^* = G/(\lambda g \rho_g(\rho_l - \rho_g))^{0.5}, \Delta T^* = C_{pl}\Delta T/h_{fg}$	P=0.1-4MPa, G=-25800 - +6250 kg/m²s, ΔT_s =1- 213K
Gambill (1963)	$q_c = q_{PB} + q_{FC}$ $q_{PB} = 0.15h_{fg}\rho_g\left(\sigma g(\rho_l - \rho_g)/\rho_g^2\right)^{0.25}$ $\left(1 + (\rho_l/\rho_g)^{0.923}C_p\Delta T_s/(25h_{fg})\right)$ $q_{FC} = 0.023k/D_e \text{Re}^{0.8} \text{Pr}^{0.4}(57\ln 145p$ $- 54p/(p + 0.103) - V/1.22 - T_b)$	D_e=3-10mm,L/D_e=6.5- 52, p=0.1-20 MPa, G= 0-5300kg/m²s
Tong and Celata (1993)	$q_c = Ch_{fg}\rho_g V/\text{Re}^{0.5}$ $C = f(p)$	D_e=2-20mm,L/D_e=100- 200, p=0.1-20MPa, G=1.3-40Mg/m²s, ΔT_s =50-150K
Gunther (1951)	$q_c = 0.072V^{0.5}\Delta T_s$	D_e=6.9mm,L/D=11, p=0.1-1.1MPa, ΔT_s=25-140K, V=1.5- 12.2m/s
Tong (1967)	$q_c = 3.155 \times \{(2.022 - 6.24 \times 10^{-2}p) + (0.1722 - 1.427 \times 10^{-2}p)$ $\times \exp[(18.177 - 0.5989p)x]\} \times [(0.1094 - 1.177x + 0.1275x|x|) \times$ $G/10^3 + 1.037] \times (1.157 - 0.869x) \times$ $[0.2664 + 0.8357\exp(-124.1D_e)]$ $\times \left[0.8258 + 0.3414 \times 10^{-6}(H_{fi} - H)\right]$ -- <u>W-3</u>	D_e=5.1-17.8mm, P=6.9-15.9Mpa, G=1356-6780kg/m²s X≤0.15 L=0.25 – 3.66m

Table 1. Subcooled flow boiling CHF correlations

3. Experiment

3.1. Technique

The onset of critical heat flux is characterized by a drastic increase in the wall temperature. For subcooled flow boiling of water with lower pressure the critical heat flux is higher and it could lead to a failure of the heated wall rapidly. This kind of CHF is also called as "fast burnout". In experiment the onset of CHF is usually detected by thermocouples for protection of the test section. However, the occurrence of CHF generally initiates from a small area (or a spot), and for a test section of larger size the CHF spot can not be expected exactly. This presents a challenge for prevention of the test section from burnout. For this condition the photocell has advantage for the detection of CHF.

In the present experiments the pressure and flow rate are kept at constant, and the CHF is approached by increasing slowly the water temperature or the power to test section. When the CHF occurs and the wall temperature exceeds about 500°C, the photocell produces an output, which switches off the power supply to test section. This technique is used by author for all the experiments in tubes and annuli.

3.2. Experimental results

3.2.1. Higher pressure CHF (Chen et al., 2011)

Experiment was performed in an uniformly heated inconel tube of 7.95 mm in diameter and 0.8 m in heating length, covering the ranges of pressure of $p = 1.96 - 20.4$ MPa, mass flux of $G = 476 - 1653$ kg/m²s, inlet subcooling of $\Delta T_{s,i} = 49 - 343$ K, outlet subcooling of $\Delta T_{s,o} = 1 - 145$ K and critical heat flux of $q_C = 0.26 - 4.95$ MW/m².

For the present low flow condition the CHF is related to the inlet condition, characterizing the mechanism of total power dominant. Totally, 193 data are obtained, and are formulated as the following empiric correlation,

$$q_C = cq_s$$

where q_s is the heat flux for the exit to reach the saturation temperature, evaluated by

$$q_s = \frac{(H_s - H_i)GD}{4L} \tag{1}$$

and

$$c = Min\left[2350(1 - 0.0307p)(G(H_s - H_i))^{-0.35}, 1.0 \right]$$

where p is the pressure, H_i and H_s the inlet enthalpy and saturation enthalpy, respectively, G the mass flux, D the diameter and L the heating length. Fig.2 shows the comparison of the prediction of Eq.(1) with the experimental data by plotting the ratio of $q_{CHF,c}/q_{CHF,M}$ versus P or G. The average deviation, AVG, is 0.75% and the standard deviation, RMS, is 5.34%.

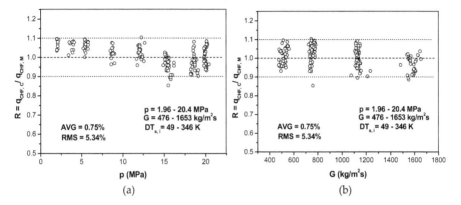

Figure 2. Comparison of experimental data of higher pressure with the prediction by Eq.(1)

The effects of mass flux, inlet subcooling and pressure are exemplified in Fig.3 and 4. The CHF decreases substantially with mass flux decreasing. As can be seen, at G > 1200 kg/m²s the data are close to the prediction of 96-CHF Look-Up Table (LUT), but at low mass flux the data are overpredicted significantly (Fig.3). For lower pressure the effects of inlet subcooling and mass flux are stronger than higher pressure, associated with complicated trend of the CHF with pressure. For G > 1200 kg/m²s higher CHF corresponds to lower pressure, especially in high subcooling region (Fig.4(a)). For G= 700 – 1200 kg/m²s the results of p < 16 MPa are not different appreciably for different pressures. For G< 600 kg/m²s, in high subcooling region lower CHF corresponds to higher pressure, while in low subcooling region lower CHF is attained at lower pressure (Fig.4 (b))

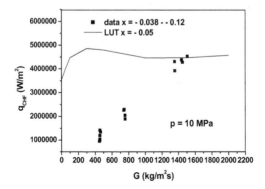

Figure 3. Effect of mass flux on the CHF

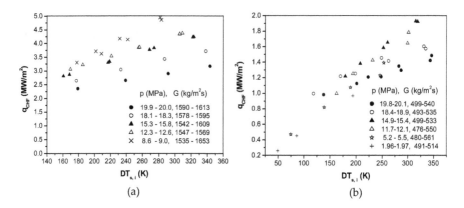

Figure 4. Variations of CHF with inlet subcooling for different mass fluxes and pressures

3.2.2. Lower pressure CHF

3.2.2.1. CHF in tubes

Medium and high subcooling (Chen et al., 2000)

Experimental data were obtained in uniformly heated stainless-steel tubes of 5.17, 8.05, 10.0 and 16.0 mm in diameter, covering the ranges of pressure p = 0.13 – 1.92 MPa, velocity V = 1.47 – 22.32 m/s, as listed in Tab.2.

Fig.5 and 6 exemplify the variations of critical heat flux with local velocity and subcooling for different pressures. The CHF increases strongly with the velocity increasing (Fig.5). For subcooling higher than about 30 K, the CHF exhibits an approximately linear increase with increase of subcooling. In medium subcooling region the CHF is not different greatly between different pressures. For p < 0.3 MPa the trend of CHF with subcooling is steeper than that of higher pressure, so that in low subcooling region lower CHF is obtained at lower pressure (Fig.6). The CHF also exhibits a general increase trend with the decrease of diameter, and at lower velocity this effect appears stronger than higher velocity.

No.	Diameter D (mm)	Length L (mm)	Pressure P (MPa)	Velocity V (m/s)	Subcooling Δ_s (K)	Number of data
1	5.17	255	0.13-1.78	3.26-22.32	6.2-89.6	62
2	8.05	383, 396	0.14-1.92	1.85-16.05	8.3 – 88.2	65
3	10.0	295, 400	0.15-1.66	3.39-9.26	30.3-89.5	53
4	16.0	295, 390	0.19-1.29	1.47-13.7	36.7-108.7	56

Table 2. The Experimental conditions for CHF in tubes

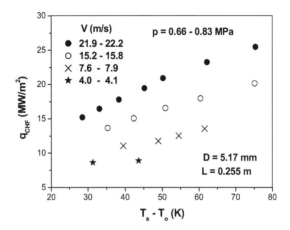

Figure 5. Effect of velocity on the CHF

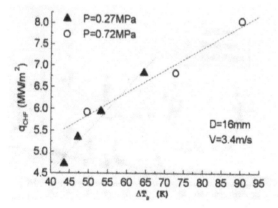

Figure 6. Effect of subcooling on the CHF

224 CHF data of subcooling higher than 35 K are formulated by the following empiric correlation with local parameters of p, V, Δ_s and D:

$$q_{CHF} = 0.109 \times 10^6 \times (1 + 0.104V) \times (15P + \Delta T_s^{1-0.1p}) \times (D/8.0)^{-0.35 - 0.05/\ln V} \tag{2}$$

where the pressure P is in MPa, velocity V in m/s, subcooling ΔT_s in K and diameter D in mm. Eq.(2) predicts the experimental data with AVG of 0.83% and RMS of 7.2%, as shown in Fig.7.

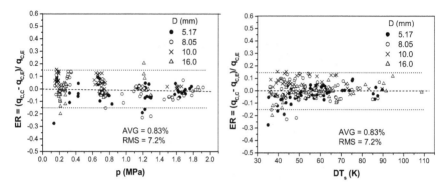

Figure 7. Comparison of experimental data of p<1.9 MPa in tubes with the prediction by Eq.(2)

Low subcooling (Chen et al., 2005).

The experiment was performed in a stainless-steel tube of D=15.9 mm with emphasis on the CHF characteristic in low subcooling region. The conditions cover the ranges of pressure of 0.2 - 1.7 MPa and velocity of 2.2 - 13.2 m/s. Fig.8 exemplifies the variations of CHF with subcooling. For p > 1.0 MPa, the CHF decreases with Δ_s decreasing monotonously. For p < 0.3 MPa, however, this trend breaks at a certain low value of subcooling, and it turns to increase at further low subcooling. The subcooling at the minimum CHF varies from 13 to 30 K with lower value corresponding to lower velocity.

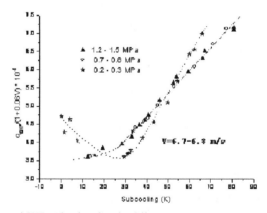

Figure 8. The variation of CHF with subcooling for different pressures

This behavior was also observed in the author's experiment in annuli and other similar experiments (Zeigarnik, 1994, Knoeble et al., 1973). It can be attributed to the onset of net vapor generation (NVG) or the onset of significant voiding (OSV). For p < 0.3 MPa, the vapor density is very small, so that in NVG regime the volumetric flow rate increases essentially, associated with a considerable increase in the liquid velocity. Therefore, the bubbles generated on the heated surface are more likely to enter into the liquid core, resulting in higher CHF. At high pressure the vapor density is much higher, and hence the CHF behavior could not vary distinctly in low subcooling region.

The NVG or OSV was generally identified by a sharper increase in the pressure drop. The following type of empiric correlation was derived

$$q^{''} = \eta \Delta T_s V^n$$

with the values of n ranging from 0.5 to 1.0. (Siman-Tov, et al. 1995). For the present experimental condition most correlations predict the OSV at subcooling of below 40 K.

Fig.9 shows the variations of pressure drop with the increase of exit temperature. As seen, before the OSV the pressure drop increases slowly with the temperature increasing. After the OSV this trend varies distinctly. Especially at low pressure it is much steeper than that at higher pressure.

In a reactor core the fuel elements are located in parallel channels. Therefore, at low pressure with low subcooling the OSV could result in flow instability or flow excursion (FE), characterized by a continuous decrease of flow rate in a channel. This could eventually lead to a failure of fuel element. Therefore, for reactor safety the limit of OSV is taken as a criterion in combination with the limit of minimum ratio of DNB

3.2.2.2. CHF in annuli

Single-side heating (Chen et al., 2004)

Experiments were performed in 8 annuli made of stainless-steel tubes with single-side heating. The diameter of outer wall is 16, 32 or 70mm and gap width is 2 to 4mm. Great majority of CHF data were obtained at outer wall and less at inner wall, covering the range of pressure of 0.17 - 1.8MPa, mass flux of 1300 to 18200 kg/m²s, outlet subcooling of 27-105 K and critical heat flux of 2.0 - 18.1 MW/m². The experimental conditions are listed in Tab.3.

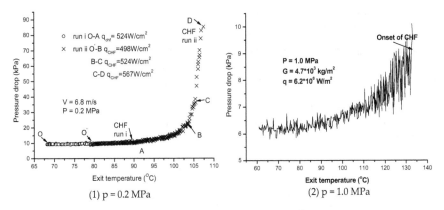

Figure 9. The variations of pressure drop with exit temperature for different pressures

No	Diameters D_2/D_1 (mm)	Length L (mm)	Pressure P (MPa)	Mass flux G (Mg/m²s)	subcooling Δ_s (K)	Number of data
1	16/12	260-400	0.17-1.31	2.8-18.2	34-89	67*
2	32/28	300	0.24-1.21	2.8-12.4	35-98	46**+27*
3	32/27.6	300	0.3-1.1	3.5-12.1	29-84	230*
4	32/26.8	275-400	0.28-1.1	2.9-13.0	37-102	67*
5	32/26	300	0.36-0.85	3.6-12.7	37-70	47*
6	32/24	300	0.18-1.8	1.3-8.0	27-105	61*
7	70/66	255	0.25-0.4	2.9-7.0	26-54	15*
8	70/65	255	0.25-0.37	3.2-5.7	34-49	18*

* outer-wall
** inner-wall

Table 3. Experimental conditions for CHF in annuli with single-side heating

The velocity and subcooling have predominant effects on the CHF. Under most conditions of interest the effect of pressure is not appreciable. The variation of gap width from 2.0 to 4.0mm does not make an appreciable effect on the CHF. This can probably be explained as follows: larger gap size associates with lower heat transfer coefficient to the liquid core, as suggested by the turbulent convection heat transfer correlation, on the other hand larger gap size corresponds to higher Reynolds number, which is benefit for the departure of bubbles from heated surface. For the present conditions these two opposite factors would be comparative, making the CHF not sensitive to the gap size. These results are consistent with many similar experiments (see Tab.1).

In the present experiment the effect of curvature of heated surface is not observed. This is understandable, because the curvature of heated surface is small, compared to the bubble, and it could not have a noticeable effect on the bubble behavior.

For the sake of simplicity in engineering applications, the effects of pressure, diameter and gap width are ignored, and the experimental data are formulated as the following empiric correlation

$$q_c = 1.21 \times 10^6 \times V^{0.5} \left(1 + 0.03 \Delta T_S\right) \tag{3}$$

The comparison of calculation of Eq.(3) with the experimental data is shown in Fig.10. The AVG and RMS are -0.01 and 0.083, respectively.

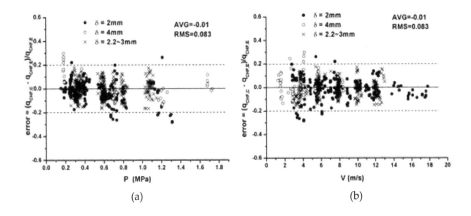

Figure 10. Comparison of the experimental data in annuli of single-side heating with the predictions of Eq.(3)

Bilateral-side heating (Chen, et al., 1996)

Experiment was performed in a stainless-steel annulus of D_1=28 mm and D_2=35.7 mm and heating length of 160 mm with bilateral-side heating. The critical heat flux data were obtained at the outer wall. The conditions cover the range of pressure of 0.31 – 0.39 MPa, velocity of 2.55 – 8.12 mm, subcooling of 49.1 – 76.6 K and the ratio of inner to outer-wall heat fluxes of q_1/q_2 of 0 – 0.94.

For convenience in comparison, the result of outer-side heating is formulated by

$$q_{c,0} = 0.96 \times 10^6 \times V^{0.43} \times (1 + 0.057 \Delta T_{s,0})$$

as shown in Fig.11.

The experimental results of bilateral-side heating are displayed in Fig.12 by plotting the ratio of $q_c/q_{c,0}$ against q_1/q_2 (1 and 2 denote the inner-wall and outer-wall, respectively). As seen, the critical heat flux exhibits an increase trend with q_1/q_2 increasing. When q_1/q_2 closes to 1.0 the CHF is increased by 15 – 20%. It can be attributed to the variation of temperature profile in liquid core, which results in an increase in condensation efficiency of the bubbly layer by subcooled liquid core, that is similar with that in single-phase convection heat transfer. This effect can be clarified further by the model analysis latter in paragraph 4.3.

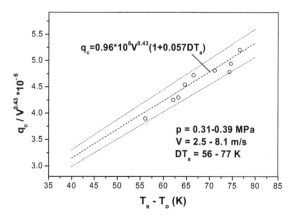

Figure 11. The CHF results of outer-wall heating

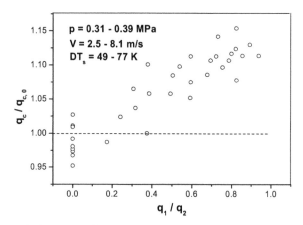

Figure 12. The ratio of $q_c/q_{c,0}$ versus q_1/q_2

3.2.3. Transient CHF (Chen et al., 2005)

Critical heat flux is more likely to occur under off-normal or accident conditions, in which a transient would experiences of flow rate, pressure and/or power. Many studies have focused on the transient CHF (Leung, 1978, Chang, et al., 1989, Iwamura, et al., 1987, 1994, Celata, et al., 1991, Weisman, 1993, Moon, et al., 2002). In higher quality region the experimental results revealed the inadequacy of the steady-state CHF correlation for transient conditions. While in subcooled and low quality region, the effect of transient on the CHF was found not appreciable. In general, the transient CHF has not been studied adequately for wider range of condition, and in evaluation of nuclear reactor safety the CHF for transient conditions is predicted with the correlations derived at steady-state conditions (IAEA-TECDOC-1203, 2001).

In the present study an experiment of flow-reduction transient CHF was performed in a stainless-steel tube of 15.9 mm in diameter, covering the range of pressure of 0.2 - 1.4 MPa, initial velocity of 4.5 - 13.5 m/s, and the initial inlet subcooling of 80 – 160 K. The flow rate was reduced linearly as

$$\dot{m} = \dot{m}_0(1 - kt)$$

where $\dot{m}$ and $\dot{m}_0$ are the instantaneous and initial flow rate, respectively, t is the time, and k is the flow reduction rate, ranging from 0.0075 to 0.24 1/s.

The experimental results are shown in Fig.13, in which the P, V and Δ_s are the instantaneous values. For P > 1.0 MPa the effect of flow transient on CHF appears not prominent. For P < 0.3 MPa, in high subcooling region the effect of transient is not appreciable, while for subcooling lower than about 50 K the result departs from the trend of steady-state distinctly, and higher CHF is attained at higher flow reduction rate.

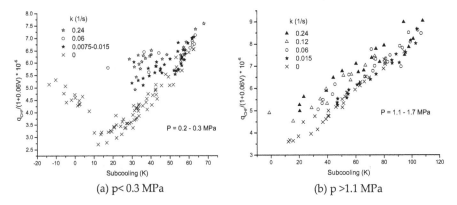

(a) p< 0.3 MPa   (b) p >1.1 MPa

Figure 13. The experimental results under flow transient condition

Different effect on the CHF observed at different conditions can probably be explained by different mechanisms of the CHF. For low subcooling the CHF is induced by a limit of enthalpy of bubbly layer, while for high subcooling the CHF is induced by a limit of bubbly-layer condensation by the subcooled liquid core (Thorogerson et al., 1974). Compared to the steady-state, at a flow-reduction transient with constant heat flux and constant subcooling the enthalpy in the bubbly layer is higher and the temperature in the liquid core is lower. At high subcooling the thickness of bubbly layer is small, thus for a constant CHF the subcooling would not be different greatly from the steady-state. While at low subcooling the thickness of bubbly layer is larger, and a constant CHF would occur at a higher subcooling than that of steady-state. It would lead to the premature of the OSV, associated with prominent effect on the CHF, as observed at low pressure.

Fig 14 shows the variation of pressure drop before the onset of CHF in flow-reduction transients of p =0.23 MPa and k = 0.24 with heat flux of q = 890, 775 and 750 W/cm² for run iii, iv and v respectively. In run iii, the pressure drop exhibits a monotonous decrease with flow rate decrease, until the onset of CHF. While in run iv and v, before the onset of CHF the decrease of pressure drop is followed by a sharp increase. In these three runs the CHF are higher than those of steady-state by about 5%, 18% and 36%. The critical subcoolings are 54.7, 46.5 and 37 K respectively, all of which are higher than the values of steady-state, evidencing the premature of OSV.

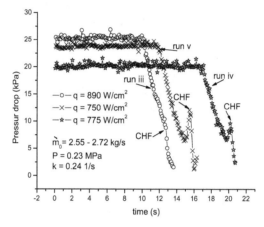

Figure 14. Variation of pressure drop before the onset of CHF

4. Physical model

4.1. Mechanism

In subcooled flow boiling the enthalpy of bubbly layer is determined by the heat transfer from the interface of bubbly layer to the liquid core, and the excessive bubble crowding serves as a thermal shield, leading to the onset of CHF. For the DNB with low subcooling or saturated condition, the critical enthalpy models were proposed by Weisman and Pei (1983) and Tong (1968), in which the heat transfer coefficient from the interface of bubbly layer to liquid core was estimated by correlations derived from their experiments. For the DNB with high subcooling, the liquid sublayer dryout models were proposed by Katto (1992), Lee and Mudawar (1983) and Celata et al. (1994), in which the bubble diameter, the thickness of liquid sublayer and the length of vapor blanket were determinant for the sublayer dryout.

For higher subcooling the major part of heat from wall is transferred to the liquid core and a minor part for increase of the enthalpy of bubbly layer. Therefore, the characteristic of bubbly layer is primarily controlled by the heat transfer from the bubbly layer to liquid core. This heat transfer is closely relative to the turbulence near the edge of bubbly layer, and is sensitive to the distance from the wall. The increase in thickness of bubbly layer has positive effect on the CHF due to increase in the heat transfer efficiency to liquid core, but negative effect due to increase in the thermal resistance of the bubbly layer. The balance of these two factors gives a critical value of the thickness. Therefore, in the liquid sublayer dryout model the thickness of bubbly layer is a determinant factor.

4.2. Model for tube (Chen et al., 2011)

The model is based on the mechanism of liquid sublayer dryout by modifying the Celata's model for the thickness of bubbly layer to cover both high and low subcooling region.

Bubbly layer

At high flow and high subcooling the minimum thickness of bubbly layer is determined by the size of a bubble, while at low subcooling it could be larger due to bubble crowding. For the present experimental condition the following expression on the thickness of bubbly layer is attempted,

$$\delta = k_1 D_B (1 + k_2 e^{-k_3 \text{Pr} Q}) \qquad (4)$$

where the factor k_1, k_2 and k_3 are the constants, Pr is the Prandtl number, Q is a parameter group (see Eq.(8)). D_B is the bubble or vapor blanket equivalent diameter, evaluated by Staub correlation (1968), as

$$D_B = \frac{32\sigma f(\beta)\rho_l}{fG^2}$$

where σ is the surface tension, ρ_l the liquid density, G the mass flux, $f(\beta)$ is a function with parameter of contact angle and recommended as $f(\beta) = 0.02\text{-}0.03$. In the present model it is taken as

$$f(\beta) = 0.03 \text{ for } p \leq 10MPa$$

and

$$f(\beta) = 0.03(1 - 0.055(p - 10)) \text{ for } p > 10MPa$$

where p is the pressure in MPa.

The friction factor, f, is calculated by Colebrook-White equation combined with Levy's rough surface model (1967), as

$$\frac{1}{\sqrt{f}} = 1.14 - 2.0\log(\frac{\varepsilon}{D} + \frac{9.35}{\text{Re}\sqrt{f}})$$

where D is the tube diameter, Re the Reynolds number, ε is the surface roughness, accounted by $\varepsilon = 0.75 D_B$.

Liquid core

The velocity distribution in the liquid core is represented by Karman's universal law, as in Celata's model,

$$U^+ = y^+ \ for \ 0 \le y^+ < 5$$

$$U^+ = 5.0 \ln y^+ - 3.05 \ for \ 5 \le y^+ < 30$$

$$U^+ = 2.5 \ln y^+ + 5.5 \ for \ y^+ \ge 30$$

with

$$U^+ = \frac{U}{U_\tau} \quad y^+ = \frac{y U_\tau \rho_l}{\mu_l}$$

and

$$U_\tau = \left(\frac{\tau_w}{\rho_l}\right)^{0.5}$$

where U is the liquid velocity, y the distance from the wall, μ_l the liquid viscosity and ρ_l the liquid density, U_τ the friction velocity, and τ_w is the wall shear stress, evaluated by

$$\tau_w = \frac{f G^2}{8 \rho_l}$$

The temperature distribution in the liquid core is as follows (Martinelli, 1947),

$$T_0 - T = Q \Pr y^+ \ for \ 0 \le y^+ < 5 \tag{5}$$

$$T_0 - T = 5Q \left\{ \Pr + \ln\left[1 + \Pr(\frac{y^+}{5} - 1)\right]\right\} \ for \ 5 \le y^+ < 30 \tag{6}$$

$$T_0 - T = 5Q \left[\Pr + \ln(1 + 5\Pr) + 0.5\ln(\frac{y^+}{30})\right] \ for \ y^+ \ge 30 \tag{7}$$

with

$$Q = \frac{q}{\rho_l C_{pL} U_\tau} \tag{8}$$

Equations (5) to (7) are assumed valid in the region of $\delta \le y \le r$, and the T_0 is a referent value, which is determined by $T = T_s$ at $y = \delta$.

Calculation of critical heat flux

The local enthalpy, H, is calculated by

$$H\dot{m} = H_C(\dot{m} - \dot{m}_{B,g} - \dot{m}_{B,l}) + H_g \dot{m}_{B,g} + H_l \dot{m}_{B,l} \tag{9}$$

where $\dot{m}$ is the total flow rate, $\dot{m}_{B,g}$ and $\dot{m}_{B,l}$ are the vapor and liquid flow rate in the bubbly layer, respectively, H_g and H_l are the vapor and liquid enthalpy, and H_C is the enthalpy of liquid core.

The $\dot{m}$, $\dot{m}_{B,g}$ and $\dot{m}_{B,l}$ are evaluated by

$$\dot{m} = \frac{\pi D^2}{4} G \tag{10}$$

$$\dot{m}_{B,g} = \pi(D-\delta)\delta\alpha_B\rho_g\overline{U_B} \tag{11}$$

and

$$\dot{m}_{B,L} = \pi(D-\delta)\delta(1-\alpha_B)\rho_l\overline{U_B} \tag{12}$$

where α_B is the void fraction in the bubbly layer, and it is taken as $\alpha_B = 0.9$, $\overline{U_B}$ is the average velocity of bubbly layer, estimated by

$$\overline{U_B} = 0.5U_{y=\delta}$$

H_C is calculated at the average temperature from the edge of bubbly layer to the center of tube, T_C, which is calculated by

$$T_C = \frac{\int_\delta^r TU(r-y)dy}{\int_\delta^r U(r-y)dy}$$

where r is the radius of tube, and δ is the distance from wall at which the temperature is equal to the saturation value.

The exit enthalpy, H, is evaluated from the heat balance equation, as

$$H = H_i + \frac{4qL}{GD} \tag{13}$$

where H_i is the inlet enthalpy, and L the heating length.

Calculation is started with a test heat flux q ($q < q_s$), and the δ, $\dot{m}_{B,g}$, $\dot{m}_{B,l}$ and T_C are calculated by Eq. (4), (10), (11) and (12). Then, the H is calculated by Eq.(9) and compared to that calculated by Eq.(13). The result of CHF is obtained through an iterative process.

To get the calculations better fit to the experimental data, the constants in Eq. (4) are as: $k_1 =$ 0.75, $k_2 = 1000$, and $k_3 = 1.0$. At low subcooling q_C is close to q_s, and not sensitive to the δ, thus the maximum value of δ is simply set as 0.1D.

The experimental data in tubes are calculated by the present model. The comparison is shown in Fig.15 and 16 for p > 2 MPa and < 2 MPa, respectively.

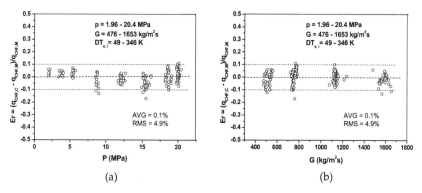

(a)   (b)

Figure 15. Comparison of the calculations of present model with the experimental results in tube for p > 2 MPa

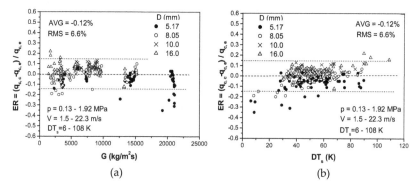

(a)   (b)

Figure 16. Comparison of the calculations of present model with the experimental results in tubes for p < 2 MPa

4.3. Model for annulus (Chen et al., 1996)

Fig.17 shows schematically the profiles of velocity and temperature in the liquid core of an annulus. Some assumptions are made as follows:

- Each wall is heated uniformly, and the flow and heat transfer conditions are fully developed;
- At the edge of liquid bubbly layer the heat flux is equal to that of outer-wall
- In liquid core the properties are evaluated at the bulk temperature.

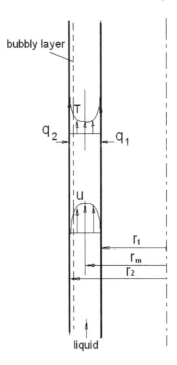

Figure 17. The profiles of velocity and temperature in the flow

In liquid core the energy balance equation is written as

$$u\frac{\partial t}{\partial x} = \frac{1}{r}\frac{\partial}{\partial r}\left[r\left(\frac{k_l}{\rho_l C_{pl}} + \frac{1}{\Pr_t}\varepsilon_m\right)\frac{\partial t}{\partial r}\right] \tag{14}$$

with

$$\frac{\partial t}{\partial x} = \frac{\overline{\partial t}}{\partial x} = \frac{2(q_1 r_1 + q_2 r_2)}{\rho_l C_{pl}\overline{u}(r_2^2 - r_1^2)} \tag{15}$$

For fully developed flow with uniform heating, the $\partial t / \partial x$ is a constant. Integrating Eq.(14) gives

$$\frac{\partial t}{\partial r} = \frac{1}{r(1+\dfrac{Pr\ \varepsilon_m}{Pr_t\ v_l})}\left[\frac{2(q_1r_1+q_2r_2)}{k_l(r_2^2-r_1^2)u}\int_{r_1}^r urdr - \frac{q_1r_1}{k_l}\right] \tag{16}$$

The average bulk temperature is approximated by

$$\bar{t} = \frac{2}{u(r_2^2-r_1^2)}\int_{r_1}^{r_i} turdr$$

Integrating Eq.(16) by parts gives

$$t_i - \bar{t} = \frac{2}{u(r_2^2-r_1^2)}\int_{t_1}^{t_i}\int_{r_1}^r turdrdt \tag{17}$$

where t_i is the temperature at the outer edge of bubbly layer. For low pressure with higher subcooling the enthalpy of bubbly layer is negligible small.

By assuming t_i equal to the saturation temperature, Eq.(17) is approximated as the subcooling Δ_s. Introducing $R=r/r_2$ and $U = u/\bar{u}$, combining of Eq.(16) with Eq.(17) gives

$$\Delta T_s = \frac{2}{1-R_1^2}\int_{R_1}^{R_i}\frac{1}{(1+\dfrac{Pr\ \varepsilon_m}{Pr_t\ v_l})R}\left[\frac{2(q_1r_1+q_2r_2)}{k_l(1-R_1^2)}(\int_{R_1}^R URdR)^2 - \frac{q_1r_1}{k_l}\int_{R_1}^R URdR\right]dR$$

The velocity distribution is assumed in power law, as

$$\frac{u}{u_m} = \left[\frac{r_j-r}{r_j-r_m}\right]^{1/5} \quad \begin{array}{l} j=1 \text{ for } r_1 < r < r_m \\ j=2 \text{ for } r_m < r < r_2 \end{array}$$

where r_1 and r_2 are the radius of inner and outer wall, respectively, and r_m is the radius of maximum velocity u_m, calculated from

$$\tau^* = \frac{\tau_2}{\tau_1} = \frac{(1-R_m^2)R_1}{R_m^2-R_1^2}$$

where τ_1 and τ_2 are the shear stress at the inner and outer wall, respectively.

The friction factor is estimated by (Xu et al., 1979)

$$f = 93\left(\frac{q}{h_{fg}\rho_g\bar{u}}\right)^{0.7}\left(\frac{h_{fg}\rho_g}{C_{pl}\rho_l\Delta T_s}\right)^{0.4}\left(\frac{\rho_g}{\rho_l}\right)^{0.2}f_0$$

with

$$f_0 = \frac{0.3164}{\text{Re}^{0.25}}$$

The momentum eddy diffusivity is evaluated by (Levy, 1967)

$$\frac{\varepsilon_m}{\nu_l} = \frac{1}{15}|(r_2 - r_m)|\frac{u^*}{\nu_l}\left(1-\eta^2\right)\left(1+2\eta^2\right)$$

with

$$\eta = (r_m - r)/(r_m - r_1)$$

and

$$u^* = \left(\frac{\tau_w}{\rho_l}\right)^{0.5}$$

The turbulent Prandtl number, Pr_t, is taken as $Pr_t = 1/1.2$.

Both the experimental data of 8 annuli with single-side heating and the annulus with bilateral-side heating are predicted by this model. The comparison is shown in Fig.18 to 20.

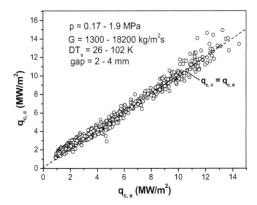

Figure 18. Comparison of the prediction of present model with the experimental data of 8 annuli

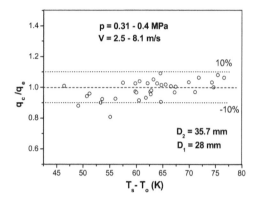

Figure 19. Comparison of the model prediction with experimental data of bilateral heating

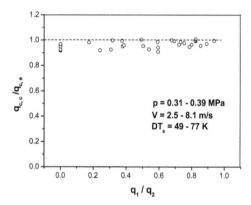

Figure 20. Effect of bilateral-side heating on the CHF by model prediction

5. Summary

Critical heat flux is an important subject to many applications. Especially for nuclear reactors, it is a major limit for the safety. The CHF has been investigated extensively over the world, and various prediction methods have been available. Unfortunately, there exists a scarcity of experimental data in certain regions. Because of extreme complexity of the phenomenon and the lack of adequate knowledge of the mechanisms, all these predictive methods are heavily relied on the experimental data, and can not be extrapolated out of the range with confidence.

In the present lab a great number of critical heat flux data of subcooled water have been obtained in tubes and annuli with different diameter and gap size over wide range of parameters with emphasis on lower pressure and higher pressure with low flow. The results fill the gap of database and the knowledge of the phenomenon.

The velocity and subcooling are the predominant parameters for the CHF. At lower pressure these effects are stronger. At the pressure below 0.3 MPa, when the subcooling decreases below a certain low value the CHF behavior varies substantially as a result of significant voiding. The effect of geometry is related to the pressure, subcooling and velocity. All these effects are inter-dependent, and are hardly to be represented in a single correlation for wide range of conditions. In the present study two models have been developed for the subcooled flow boiling CHF in circular tube and annulus, based on the mechanisms of CHF and the present experimental data. They will be validated and improved for extended range of conditions.

Nomenclature

A	flow area
A_H	heated area
C_p	specific heat
D	diameter
Δ_s	subcooling
f	friction factor
G	mass flux
h	heat transfer coefficient
h_{fg}	latent heat
H	enthalpy
k	thermal conductivity
L	heating length
p	pressure
q	heat flux
r	radius
T,t	temperature
U u	velocity in axial direction
V	average liquid velocity at CHF
x	quality
Nu	Nusselt number
Pr	Prandtl number
Pr_t	turbulent Prandtl numbe
Re	Reynolds number
W	channel width
y	distance from the wall
ρ	density

σ	surface tension
ε	surface roughness
ε_m	momentum eddy diffusivity
μ	dynamic viscosity
ν	kinematic viscosity
τ_w	wall shear stress
α_B	void fraction
λ	critical wavelength

Subscript

B	bubble
c	critical, calculatiom
l	water
g	steam
m	maximum velocity
M	measurement
s	saturation
1	inner-wall
2	outer-wall

Author details

Yuzhou Chen
China Institute of Atomic Energy, China

6. References

Berglest, A. E. (1977), Burnout in Boiling Heat Transfer, Part II: Subcooled and Low-Quality Forced-Convection System, *Nuclear Safety*, Vol. 18, N2, pp 154 - 167

Chang, S. H., Lee, K. W. and Groeneveld, D. C. (1989), Transient-Effects Modelling of Critical Heat Flux, *Nucl. Eng. Design*, Vol. 113, PP 51 - 57

Celata, G. P., Cumo, M., D'annibale, F., Farello, G. E. and Mariani, A. (1991), CHF Behavior During Pressure, Power and/or Flow Rate Simultaneous Variations, *Int. J. Heat Mass Transfer*, 34(3). PP 723 - 738

Celata, G. P., Cuma, M., Mariani, A. Simosini, M. and Zummo, G. (1994), Rationalization of Existing Mechanistic Models for the Prediction of Water Subcooled Flow Boiling Critical Heat Flux, *Int. J. Heat and Mass Transfer* Vol. 37, Suppl. 1, PP 347 - 360

Chen, Y., Zhou, R. and Chen, H. (1996), Critical Heat Flux for Subcooled Flow in an Annulus with Bilateral Heating, in *Heat Transfer and Technology 1996*, Bu-xuan Wang, Higher Education. Press, ISBN 7-04-005894-4, pp 376 - 381

Chen, Y, Zhou. R., Hao L. and Chen, H. (1997), Critical Heat Flux with Subcooled Boiling of Water at Low Pressure, *Proc. 8th Int. Topical Meeting on Nuclear Reactor Thermal-Hydraulics*, 1997, Japan, Vol. 2, PP 958 - 964

Chen, Y., Zhang, H., Guo F. and Hao L. (2000), Subcooled Flow Boiling CHF in Tubes with Different Diameters, Presented at *Boiling 2000*, Alaska

Chen, Y., Zou, L. and Yang, C. (2004), Subcooled Critical Heat Flux in Annuli at Lower Pressure, *Proc. NUTHOS-5*, Japan

Chen, Y., Yang, C. and Mao, Y. (2005), An Experimental Study of Subcooled Flow Boiling Critical Heat Flux of Water under Steady-State and Flow-Transient Conditions at Lower Pressure, *Proc. 11th Int. Topical Meeting on Nuclear Reactor Thermal-Hydraulics (NURETH-11)*, Avignon, France

Chen, Y., Yang, C., Zhao, M., Bi, K., Du, K. and Zhang, S. (2011), Subcooled Boiling Critical Heat Flux of Water Flowing Upward in a Tube for Lower Flow and Pressure up to 20 MPa, *Proc. 14th Int. Topical Meeting on Nuclear Reactor Thermal Hydraulics (NURETH-14)*, LN-620

Gambill, W. R. (1963), Generalized Prediction of Burnout Heat Flux for Flowing Subcooled Wetting Liquids, *Chem. Eng. Prog. Sump. Series*, Vol.59, No.41

Groeneveld, D. C., et al. (1996), The 1996 Look-up Table for Critical Heat Flux in Tubes, *Nuclear Eng. Design* Vol. 163, PP 1 - 23

Gunther, F. C. (1951), Photographic Study of Surface Boiling Heat Transfer to Water with Forced Convection, *Trans. ASME*, Vol.73, PP 115

Iwamura, T. (1987), Transient Burnout Under Rapid Flow Reduction Condition, *J. Nucl. Sci. and Tech.* Vol.24(10), PP 811 - 820

Iwamura, T., Watanabe, H. and Murao, Y. (1994), Critical Heat Flux Experiments under Steady-state and Transient Conditions and Visualization of CHF Phenomenon with Neutron Radiography, *Nuclear Engineering and Design* Vol. 149, PP 195 - 206

Katto, Y. (1992), A Prediction Model of Subcooled Water Flow Boiling CHF for Pressures in the region 0.1-20.0 MPa, *Int. J. Heat ans Mass Transfer*, Vol. 35, PP 1115 - 1123.

Klausner, J. F., Mei, R., Bernhard, D. M. and Zeng, L. Z. (1993), Vapor Bubble Departure in Forced Convection Boiling, *Int. J. Heat Mass Transfer*, Vol. 36(3) PP 651 - 662.

Knoebel, D.H., Harris, S.D. and Crain, B. (1973), Forced Convection Subcooled Critical Heat Flux, D_2O and H_2O Coolant with Aluminum and Stainless-Steel Heaters, *DP - 1306*

Levy, S. (1967), Turbulent Flow in an Annulus, *Trans. ASME, C, J. Heat Transfer*, Vol. 89 (1), PP 25 - 31

Lee, C. H. and Mudawar, I. (1983) A Mechanistic Critical Heat Flux Model for Subcooled Flow Boiling Based on Local Bulk Flow Conditions", Int. J. Multiphase Flow, Vol. 14. PP 711 - 728

Leung, J. C. M. (1978), "CHF under Transient Conditions: a Literature survey," *NURETH/CP-0056*

Martinelli, R. C. (1947) "Heat Transfer to Molter Metals", Trans. ASME Vol. 69, pp 947 – 951

Mirshak, S., Durant, W. S. and Towell, R.H. (1959), Heat Flux at Burnout, *DP - 355*

Moon, S. K., Chun, S. Y., Choi, K. Y. and Baek, W. P. (2002), Transient Critical Heat Flux Under Flow Coastdown in a Vertical Annulus With Non-Uniform Heat Flux Distribution, *J. Korea Nuclear Science*, Vol. 34(4) PP 382 - 395

Siman-Tov, M., Felde, D. K., McDuffee, J. L. and Yoder. Jr. G. L. (1995), Static Flow-instability in Subcooled Flow Boiling in Wide Rectangular Parallel Channels, 2nd *Int. Conf. on Multi-phase Flow*, Japan

Staub, F. W. (1968), "The Void Fraction in Subcooled Boiling – Prediction of the initial Point of Net Vapor Generation, *J. Heat Transfer*, Vol. 90, PP 151 - 157

Sudo, Y. and Kaminaga, M. (1993), A New CHF Correlation Scheme Proposed for Vertical Rectangular Channels Heated From Both Sides in Nuclear Research Reactors. *ASME, C.* Vol. 115, PP 426 - 434

Thermohydraulic relationships for advanced water cooled reactors (2001), *IAEA-TECDOC-1203*

Thorogerson, E. J., Knoebel, D. H., and Gibbons, J. H. (1974), "A Model to Predict Convective Subcooled Critical Heat Flux," Trans. ASME, J. Heat Transfer, 96, 79-82

Tong, L. S. (1968), Boundary Layer Analysis of the Flow Boiling Crisis, *Int. J. Heat Mass Transfer*, Vol. 11, PP 1208 – 1211

Tong, L. S. (1967), Prediction of Departure from Nucleate Boiling for An Axially Non-Uniform Heat Distribution, Journal Nuclear Energy, Vol. 21

Tong, L. S. and Tang, Y. S. (1997), *Boiling Heat Transfer and Two-Phase Flow*, Taylor & Francis Pub., ISBN 1-56032-485-6

Weisman, J. and Pei, B. S. (1983), Prediction of Critical Heat Flux in Flow Boiling at low qualities, *Int. J. Heat Mass Transfer*, Vol. 26, PP 1463 – 1477

Weisman, J. (1995), A Phenomenological Explanation of the Relationship Between Steady State and Transient CHF at Subcooled or low Quality Conditions, *Nucl. Eng. and Des.* Vol. 158, PP 157 - 160

Xu, G. at al. (1979), Heat Transfer and Pressure Drop in an Annulus with Subcooled Boiling Water, Analysis and Experiment of Reactor Thermohydraulics, Atomic Energy Press, PP 412 – 417 (in Chinese)

Zeigarnik, Yu. A. (1994), Critical Heat Flux with Boiling of Subcooled Water in Rectangular Channels with One Sided Supply of Heat, *Thermal Engineering*, Vol. 28 PP 40

Droplet Impact and Evaporation on Nanotextured Surface for High Efficient Spray Cooling

Cheng Lin

Additional information is available at the end of the chapter

1. Introduction

Resulting from Moore's law in semiconductor technology, the progresses such as shrinking feature size, increasing transistor density, and improving circuit speeds, lead to higher chip power dissipations and heat fluxes. Consequently, new and novel cooling techniques are of interest. Bar-Cohen et al. [1] have reviewed several techniques for direct liquid cooling, such as pool boiling, gas-assisted evaporative cooling, jet impingement, spray cooling and synthetic jets, emphasizing the important implications of a direct liquid approach in the thermal management of hot spots, where heat fluxes can be as high as 1~2 kW/cm². Nanotubes have unique properties as discussed by Berber et al. [2] are reported to have measured high thermal conductivities around 6600 W/mK at room temperature for carbon nanotubes. These can be placed in the thermal interface material to provide a low heat resistance path through the thermal interface material, significantly improving the thermal conductivity of the TIM. Two-phase heat transfer involving the evaporation of a liquid in a hot region and the condensation of the resulting vapor in a cooler region can provide large heat fluxes needed for microelectronic packages to operate at acceptable temperature levels. Spray cooling, which involves the boiling of a working fluid on a heated surface, is an example of efficient heat transfer scheme that exploit the benefits of two-phase heat transfer. Nanotextured surfaces provide new opportunities to improve the controllable fluid and heat transport for thin film evaporation. Significant efforts in previous work focused on spray cooling of microstructured surfaces [3,4] and the employment of nano-textured surfaces to achieve an enhanced boiling heat transfer [5]. Carbon nanotubes (CNTs) forests have more recently been investigated to enhance nucleate boiling and film boiling [6]. Four regimes have been identified: flooded, thin film, partial dry-out and dry-out. The heated surfaces of micro-structure are most suited in the thin film and partial dry-out regimes because of the wetting enhancement. Sodtke and Stephan [7] demonstrated that micro-structured surfaces lead to an increased contact line length and thereby increase the overall heat flux. The proposed enhanced boiling mechanism is the integration of spray cooling on nanotextured surfaces, which is expected to improve the heat transfer coefficient over 10 times (up to 1000 W/cm²). Numerous studies have been proposed cooling methods using spray cooling and pooling boiling with nanostructure have achieved very high heat fluxes; they are listed in Table 1 [8-13]. Amon et al.[8] and Hsieh and Yao[9] studied the heat transfer on square microstuds (160~480 mm size, groove depths 333~455 μm, groove widths 120~360 μm) manufactured on silicon. Two full-cone pressurized spray nozzles (60° and 80° cone angles, flow rates up to 4.41 g/cm² min, d$_{32}$ between 75 and 100 μm) were used to spray water at very low flow rates onto the surfaces. Surface texture was found to have little effect in the single-phase and dryout regimes. The authors attributed the higher heat transfer observed for the microtextured silicon surfaces in the intermediate regimes to more effective spreading of the liquid by capillary forces. A plain aluminum surface was found to have higher heat transfer than a silicon surface, but this disadvantage could be overcome by surface texturing. The maximum heat flux achieved was just over 50 W/cm². Visaria and Mudawar [10] developed a relatively new universal CHF correlation that combines the spray inclination functions and corrected subcooling constant. The CNT coating, the parallel vertical CNTs as well as the mesh of CNTs create deep, near-zero-angle cavities that are ideal for embryo formation, especially for the low contact angle coolants. Ujereh et al.[11] also demonstrated significant CHF enhancement with CNTs for both silicon and copper walls because of the aforementioned increase in surface area. Li et al. [12] fabricated Cu nanorods having 50 nm diameter and 450 nm height using oblique-angle deposition. In a pool boiling experiment of water, the wall superheat was decreased, and the CHF was improved by about 10% compared to a flat Cu surface. The authors showed that multiple scales from nano to micro play a key role in enhancing the nucleate boiling performance. Chen et al. [13] synthesized Cu and Si nanowires by electroplating Cu into nanoscale pores and aqueous electroless etching (EE) techniques respectively. The pool boiling experiments with water showed more than 100% increase in the CHF value. The present work focuses on pool boiling of saturated water on nanowires, made of Si or Cu, and observe significant enhancement of both critical heat flux and thermal conductance on nanowires compared to plain surface. The reported CHF (~200 W/cm²) is among the highest values for pool boiling heat transfer.

The remarkable boiling heat transfer performance of the proposed scheme is mainly attributed to the following four characteristics: (1) the large latent heat of vaporization which makes boiling a very efficient mode of heat transfer, (2) the nanostructures which achieve higher cooling performance in the thin film and partial dry-out regime, because more water will be retained on the heat transfer surface due to the capillary force, (3) the enhancement in critical heat flux (CHF) which is realized by the increase of boiling area on nano- textured surfaces, and (4) the super-hydrophilic nano-textured surface which is expected to result in much higher evaporative heat transfer rate. Recently, we have demonstrated manipulation schemes that can passively drive water droplets undergoing spontaneous self-directed motion upon contact with a chemically patterned nano-textured surface (nano-wetting effect) and with a surface tension gradient [14,15]. As such, small droplets can be transported at high rates to quickly remove dissipated heat from the surface. Following the presentation of a recent conference paper [16], this work proposes enhanced boiling mechanism is the integration of spray cooling on nanotextured surfaces, which is expected to improve the heat transfer coefficient over multiple times.

Reference	Working fluid	Droplet size	Heater size	Dominant heat transfer mechanism	Critical Heat Flux (CHF)
2005 ASME JHT (Amon et al.) [8]	HFE-720	50~100 μm	25.2×25.2 mm^2	Spray cooling (Evaporation and boiling)	45 W/cm^2
2006 IJHMT (Hsieh et al.) [9]	Water	75~100 μm	25.2×25.2 mm^2	Spray cooling (Evaporation and boiling)	50 W/cm^2
2009 IEEE CPAT (Visaria and Mudawar) [10]	Water, FC-72, FC-77, FC-87 and PF-5052	111~249 μm	10×10 mm^2	Spray cooling (Evaporation and boiling)	100 W/cm^2
		Nanowire size/ Contact angle(CA)			
2007 IJHMT (Ujereh et al.) [11]	FC-72	50 nm(dia.) 20~30 μm(height) Very small CA	12.7×12.7 mm^2	Pool boiling (Nucleate Boiling)	~30 W/cm^2
2008 Small (Li et al.) [12]	Water	40~50 nm (dia.) 450 nm(height) CA=38.5 °	10×10 mm^2	Pool boiling (Nucleate Boiling)	~160 W/cm^2
2009 Nano Letter (Chen et al.) [13]	Water	200 nm(dia.) 40~50 μm(height) CA~0 °	10×10 mm^2	Pool boiling (Nucleate Boiling)	~200 W/cm^2

Table 1. Summary of previous works on evaporation/boiling on micro/nano-structure coated surface

2. Experiment set-up and procedure

Deionized water impinging on solid surface was studied by recording the impingement process with a high speed camera at 2000 frames/s. A schematic of the experiment set-up is shown in figure 1. The component include the following: droplet generation system, nano-textured surface and heating system, high-speed charged coupled device (CCD) camera (Ultima APX Fastcam, Photron Ltd., Japan) and lighting system. All components were synchronized so as to achieve simultaneous droplet imaging and thermal measurements.

The water droplets used in the experiment had diameter of ~2 mm and velocity of ~0.3 m/s. Three different impingement surfaces were used: an oxidized silicon, a vertical CNTs and curved CNTs surface. The range of diameters and lengths of the CNTs is about 57~105 nm and 4.28~8.48 μm, respectively. In table 2, the detail specification of the growth of CNTs on the silicon substrate, and test section are presented.

MWCNTs have been synthesized on Ni/Ti multilayered metal catalysts by thermal CVD at 400 °C, in which acetylene was used as carbon source in a 3-inch diameter quartz tube of the furnace. The Ni/Ti multilayered metal catalysts were found to decrease CNT growth temperature effectively. We used a temperature ramp to 800 ⊚ at a rate of 20 ⊚/min with a constant argon flow rate of 400 sccm (standard cubic centimeter per minute). At the stable temperature, followed by Ar (200 sccm) and NH_3 (200 sccm) were added to the furnace for 10 minutes. Finally, the chip cooled down at a rate 5 °C/ min. The coating thickness and nanowire size were measured with a micrometer and shown in an SEM images. Figure 2 shows the cross sectional view of the carbon nanotubes coated surface. The as-grown CNTs are superhydrophobic with a measured contact angle as high as 146°. The carbon nanotubes were functionalized using two methods: (1) H_2O plasma was utilized during the treatment, the flow rate of H_2O plasma was set at 1 sccm with duration of 30 sec.(2)aqueous solution of 6M H_2SO_4 at 80°C for 1 hour as shown in Figure 3. Water contact angle measurement using a FTA200 system (First Ten Angstroms Inc.) indicated the contact angle of 4.7°±1.0° (superhydrophilic) for the vertical CNTs surfaces and the contact angle of 1.1°±0.3° (superhydrophobic) for the curved CNTs surfaces. This functionalization of CNTs is superhydrophilic with a measured contact angle as low as 5° are shown in Figure 4.

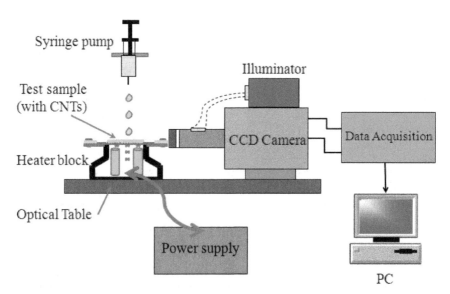

Figure 1. Schematic of an experimental apparatus.

Type Parameter	Vertical CNTs	Curved CNTs
Height h(μm)	7.28~8.48	4.28~7.14
Diameter D(nm)	70.9~92.1	57.3~105.4
Silicon substrate W×L×H (mm)	10×10 ×0.4	10×10×0.4

Table 2. The structural parameter of test sections

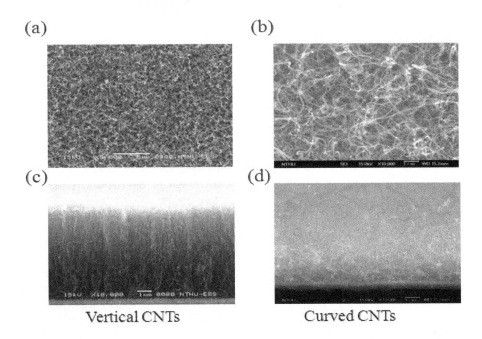

Figure 2. SEM images of the different testing surfaces:(a) top view of the vertical CNTs,(b) top view of the curved CNTs, (c)side view of the vertical CNTs, (d) side view of the curved CNTs.

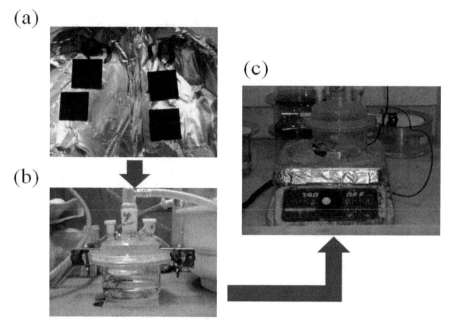

Figure 3. Schematic of the samples with CNTs surface treatment sequence: (a) top view of the CNTs samples,(b) vacuum for 5minutes using pump, (c) 80 °C, H_2SO_4 solution for 1hr.

The heat flux and the heat transfer coefficient of the heated surface were evaluated with the temperature difference (Δ), by measuring the mean temperature between the surrounding temperature around the heated surface and the wall temperatures read by the thermocouples in the copper block. The heat flux is obtained as follows,

$$q'' = \frac{IV}{A} = h(T_s - T_a) = h\Delta T \tag{1}$$

Note that the heat transfer coefficient in Eq. 1 can also be written in the form

$$h = \frac{q''}{(T_s - T_a)} \tag{2}$$

where q″ is the heat flux (W/cm^2), I is the current (Ampere), V is the voltages, and A is the heating surface area of the heated surface. The total amount of heat supplied to the heater was measured by using a power meter system, and the data for all temperatures and powers were recorded by using a data acquisition system. Heat flux was regulated successively by changing the voltage input through the DC power supply. When steady state was reached (the variation of Δ in the range of approximately 0.2 °C), all required data were stored, and the next step was executed at a higher voltage.

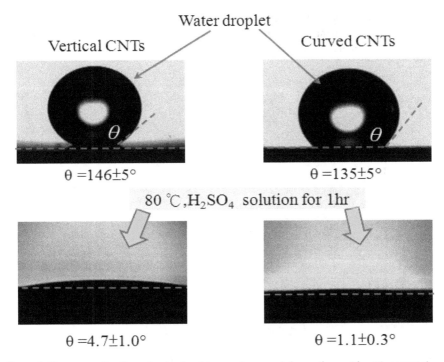

Figure 4. The contact situations of water droplets on carbon nanotubes surfaces without (upper row) and with (bottom row) H₂SO₄ treatment

3. Theoretical model

One aspect of spray cooling is basic boiling phenomena, which should be considered as a prerequisite to the more complex process of spray cooling. Therefore, consider boiling in the absence of an incoming spray. Boiling is a highly efficient means of heat transport in which liquid is vaporized due to the temperature of the liquid exceeding the saturation vapor pressure. In boiling, heat can be removed by increasing the temperature of the liquid (sensible heating) as well as the vaporization process (latent heating). As shown in Figure 5 [17], the boiling curve is a plot of surface heat flux versus excess surface temperature above saturation. When a preheated alloy exits the die in an extrusion, forging, or continuous casting process, it is typically at a temperature above the Leidenfrost point (D point) and the surface experiences film boiling. This boiling regime is characterized by a thermally-insulating layer of steam forming between the surface and individual impinging drops, resulting in poor heat transfer. When the Leidenfrost temperature is reached, the vapor film is interrupted by partial contact of liquid with the surface, causing the surface heat flux to increase with decreasing temperature until the point of critical heat flux (C point: CHF). After CHF, the surface is cooled by nucleate boiling until the temperature falls into the single-phase regime. Spray cooling is preferred to quenching in stagnant liquid because it raises the Leidenfrost temperature and enhances significantly the heat transfer rate even in the film boiling regime.

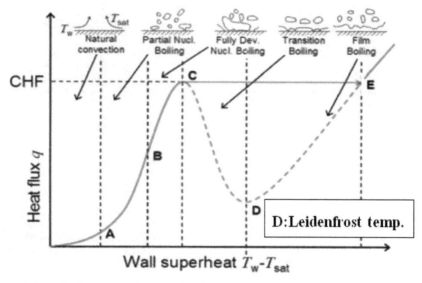

Figure 5. Typical boiling curve and associated boiling regimes showing heat flux vs. wall superheat (T_w - T_{sat}) (adapted from ref. 17)

3.1. Impinging droplets on wetting surface modeling

The dynamics of droplet impact on the two types CNTs surfaces modified by aqueous solution of 6M H_2SO_4 at 80°C for 1 hour were characterized to demonstrate how the surface energy influences drop impact transitions from bouncing to wetting. The antiwetting pressure (P_A) and wetting pressure (P_w) influenced the wetting states of impinging droplets. When Pw was larger than Pa, the droplet struck the surface in a wetting state. The wetting pressure (Pw) is given by

$$P_W = \frac{1}{2}\rho V^2$$

(3)

where ϱ and V are the density and the velocity of the impinging liquid, respectively. For nanotextured surface, the antiwetting pressure is capillary pressure. The maximum value of P_A is calculated as the Laplace pressure of the maximum deformation of the water-air interface between posts of nanostructures, the capillary pressure is defined as

$$P_A = -2\sqrt{2}\gamma_{lv}\cos\theta_A / D$$

(4)

where γ_{lv} is the surface energy of the water at the liquid-vapor interface (0.073 N/m), θ_A is the advancing CA of the water droplet on the flat surface, and D is the spacing between the nanoscale posts of the nanostructure surface. However, this model has limitations in that only two wetting states are available: bouncing or wetting. Thus, for example, the wetting state of a droplet partially pinned at the contact area cannot be sufficiently explained by these two pressures. A new wetting pressure, the effective water hammer pressure(P_e) at the contact stage, is therefore introduced. At the moment of impact on the nanostructure, the compressed region of the droplet with pressure (P_e) pushes on the liquid-air interface between the nanoposts and induces the penetration of liquid in the contact area. The maximum possible P_e for a spherical droplet impinging at low impact speed is defined by

$$P_e \approx 0.2\rho CV \tag{5}$$

where C is the speed of sound in water (1497 m/s). When P_A is smaller than Pe but larger than Pw, the droplet is in a partial wetting state. The water droplets used in the experiment had diameters of ~2 mm and velocity of ~0.3 m/s. A high-speed camera (up to 4 K frames/s) was used in the study. Relative magnitude of the wetting and antiwetting pressures decides the wetting states of impinging droplets: (1) total wetting state (P_e>P_w>P_A) as water penetrates in both contact and spreading stage.(2) partial wetting state(P_e> P_A >P_w) as water penetrates only during contact stage.(3) Total nonwetting state (P_A > P_e> P_w) as the structure resist wetting in both stages [18]. In this experiment the droplet completely wetting state within the modified CNTs surfaces, with penetration of water in both contact and spreading stages.

3.2. Key parameter of evaporative spray cooling

The surface enhancement is realized by an array of square pins. The microstructures on the heating surface not only increase the heat transfer area but also provide a driving force for liquid spreading. To characterize the capillary force between micro-pins, a dimensionless parameter, Bond number (Bo), be expressed as

$$Bo = \frac{G}{\sqrt{\gamma / (\rho_l - \rho_v)g}} \tag{6}$$

The Bond number was defined as the ratio of the gravitational force to the surface tension force. Generally, narrower grooves (smaller G; smaller Bo) are more desirable in evaporative heat transfer because of the better liquid spreading ability. Hence, there may exist on an optimal groove width for the evaporative water spray cooling on enhanced silicon surfaces. Another influential surface factor in current study is the bottom surface area of the grooves, because it is directly related to the surface area available for the liquid evaporation in the grooves.

Several control parameters that influence on the impact dynamics include: droplet size, liquid viscosity μ, and impact velocity V_0 .We describe these effects in terms of dimensionless numbers: the Weber number We_d, the ratio of kinetic energy to surface energy, characterizing the deformability of the droplet; the Reynolds number Re, the ratio of inertia to viscosity effect:

$$We_d = \frac{\rho_f V_0^2 d}{\sigma}; \ Re_d = \frac{\rho_f V_0 d}{\mu} \tag{7}$$

Here d the diameter of the liquid drop, V_0 is the impact velocity, ϱ is the liquid density, σ is the surface tension, and μ is the liquid viscosity.

At wetting heat transfer, the droplets can be in continuous or semi-continuous direct contact with heat surface. After an initial period of transient conduction heat transfer, the droplets enter into either nucleate or transition boiling regimes. In this case, the droplet incoming Weber number may have a week effect on enhancing the droplet breakup. Wet (hydrophilic surface) cooling results in a significant drop in the surface temperature due to its highly cooling efficiency. In non-wet cooling, also referred to as film boiling, a significant amount of water vapor is generated between the heat surface and the droplet, thus preventing direct contact. Since vapor has a very low thermal conductivity, it acted as insulation between the surface and the incoming spray; therefore, lower cooling efficiency. In this boiling regime, the impact droplet velocity (or Weber number) has a significant influence on the cooling efficiency. For low Weber number, droplets cannot penetrate through the film layer. For high Weber number, droplets can penetrate through the film layer, and more surface contact can be established. The impact velocity for this case is ~30 cm/s and the corresponding Weber number (We_d) is 2.47.

3.3. Empirical model of evaporative spray cooling

In spray cooling, empirical models have been developed with the continuous expansion of experimental data based on system of interests. Mudawar and Estes (19) first attempted an empirical model to predict CHF in spray cooling by correlating CHF with the volumetric flux of liquid and the Sauter Mean Diameter of droplets, as following:

$$\frac{q_{CHF}}{\rho_v h_{lv} V} = 1.467 \left[(1 + \cos(\frac{\theta}{2}))\cos(\frac{\theta}{2}) \right]^{0.3} \left[\frac{\rho_l}{\rho_v} \right]^{0.3} \left[\frac{\rho_l V^2 d_{32}}{\sigma} \right]^{-0.35} \left[1 + 0.0019 \frac{\rho_l C_{pl} \Delta T_{sub}}{\rho_v h_{lv}} \right] \tag{8}$$

where θ is the spray cone angle, d_{32} is the Sauter Mean Diameter, σ is the surface tension, $\Delta T_{sub} = T_{sat} - T_l$ is liquid subcooling at nozzle inlet, V is average volumetric flux over spray impact area, and h_{lv} is the evaporative latent heat. To predict CHF using Eq.(8), the nozzle parameters and droplet parameter (pressure drop across the nozzle, volumetric flow rate, inclined angle, and Sauter Mean Diameter of droplets) have to be tested. In addition, the distance between the nozzle orifice and the heated surface needs to be chosen carefully, so that the spray cone exactly covers the heated surface. This model was validated by a set of experiments of the spray cooling on a rectangular 12.7×12.7 mm² flat surface using refrigerants (FC-72 and FC-87). The volumetric flow rate was regulated inside the range of $(16.6{\sim}216) \times 10^3$ m³ · s⁻¹/m². The Sauter Mean Diameter of droplets was inside the range of 110~195 μm. The superheat temperature was below 33 °C. The accuracy of this model was claimed to be within ±30%.

Fluid flow for droplet impact was modeled using a finite-difference solution of the Navier–Stokes equations in a three-dimensional Cartesian coordinate system. The liquid is assumed to be incompressible and any effect of the ambient air on droplet impact dynamics is neglected. The fluid flow is assumed to be laminar. The free surface of the deforming droplet is tracked using the volume of fluid (VOF) method which solves the time-dependent Navier-Stokes equations, and interface evolution equation to simulate a mixture of two immiscible fluids with surface tension described in [20].

4. Results and discussion

During the entire impingement process, the droplet goes through impact, advancing, maximum spreading, receding and a period of oscillation with pinned contact line. For the non-heating (T_s=24°C) case once the oscillation decrease the drop gradual equilibrium at a 55°contact angle on oxidized silicon surface in Figure 6(a). Figure 6(b) shows time sequence images of a water droplet falling on superhydrophilic surface. Figure 6 indicates the different droplet dynamics variables. From the figure 7 prior to impact, the droplet diameter (d_0) and impact velocity (V_0) were measured. The time is determined to be zero at the moment of impact and the spreading time can be calculated relative to the impact time.

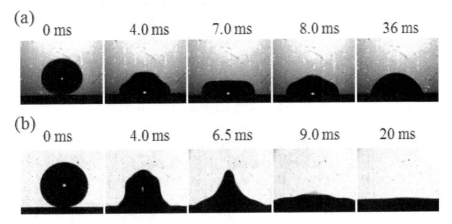

Figure 6. Illustration of droplet dynamic impinging process : (a)on the oxidized flat surface, (b)on hydrophilic curved-CNTs surface.

Images of a typical impingement with boiling process are shown in figure 8, where (a)-(e) are the images corresponding to the five stages listed. During initial impact, the droplet rapidly spreads, rebounds and then oscillates at a near-constant wetting diameter. Boiling involves initial nucleation followed by severe boiling including droplet ejection and expansion, and evaporative cooling with the absence of boiling. At non-boiling heating levels the droplet experiences evaporation once the oscillation period ends. Droplet height decreases while the wetting diameter remains nearly constant. When a critical angle is reached, the droplet diameter diminishes rapidly and then dry-out occurs.

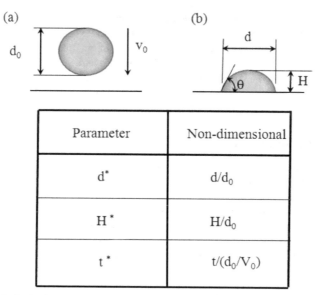

Figure 7. Droplet hydrodynamics parameters: (a) pre-impact, (b) post-impact

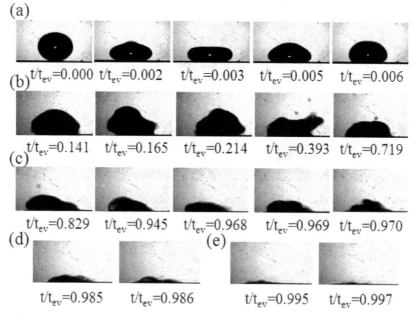

Figure 8. Droplet images of the five stages of impingement process for water impinging on the curved CNTs surface at Ts = 187°C: (a) initial impact, (b) boiling, (c) near-constant wetting diameter evaporation, (d) fast receding contact line evaporation, (e) final dry-out

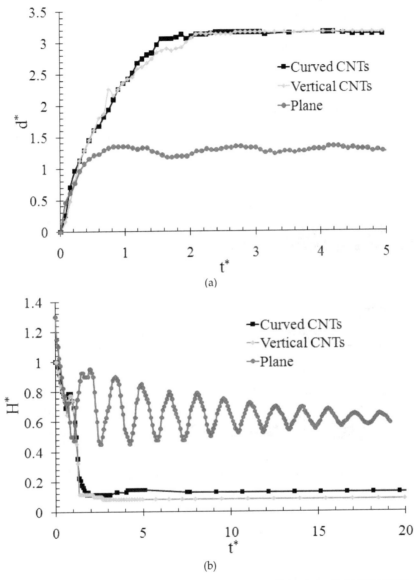

Figure 9. Initial impact non-dimensional parameter time evolution illustration at $T_s=24$ °C:(a)d*[d/d$_0$] versus t*[t/(d$_0$/V$_0$)],(b) H*[H/d$_0$] versus t*[t/(d$_0$/V$_0$)]

Figure 9(a) shows a rapid initial advancing to a maximum diameter followed by a rebound. The maximum spreading for all cases occurs within t*<2 at T_s = 24°C. Figure 9(b) show that the droplet height at essentially the same phase oscillating at a 55°contact angle on oxidized silicon surface, where a higher droplet height results in larger contact angles. These results indicate that the surface energy differences are enhanced somewhat for the nanotextured surface with a reduction of the relative advancing and receding contact angles during the initial impact. Also, the contact angle oscillations that occur once spreading has come to equilibrium seem to be damped. Image of water impingement on three surfaces with boiling are shown in figure 10, where (a)-(c) are the images corresponding to the full-drop bouncing back. The rebounding surface temperature of oxidized silicon, vertical CNTs and curved CNTs comparison are 165°C, 247°C and 257°C, respectively, for surface temperatures at a heating (T_s) and rebounding (T_r) condition.

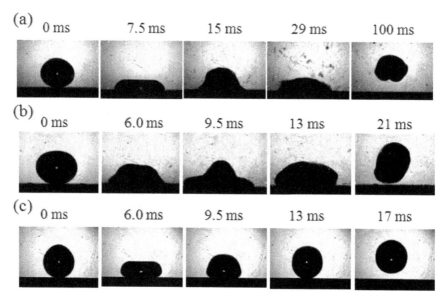

Figure 10. Snapshots of free fall droplets released from various surfaces at rebounding temperature (T_r):(a) on the oxidized silicon at T_r=165°C,(b) on the vertical CNTs surface at T_r=247°C, (c)on the curved CNTs surface at T_r=257°C .

Results show that the curved wire nanostructured surface enhanced capillary pumping effect and retained water at higher surface temperatures (T_s=226°C), as seen in figure 11. During the advancing phase the contact line length is nearly identical for the two nano-textured surfaces, however, during the retraction phase the contact line for the droplet on the curved CNTs at a surface temperature of 248°C becomes pinned, while the droplet on the vertical CNTs at a surface temperature of 248°C retracts and then rebounds off the surface is shown in figure 12. The increased spreading area over the curved CNTs surface suggests a greater potential for increased liquid lateral transportation and retention, thus bring out the consequence in improving high efficiency cooling.

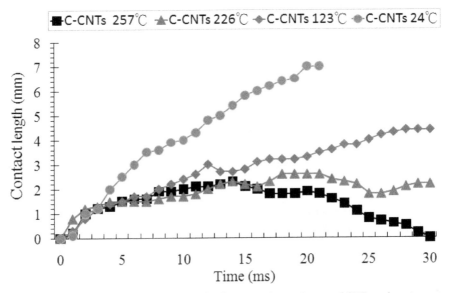

Figure 11. Contact line length versus time for droplets impinging on the curved CNTs surface at different temperatures

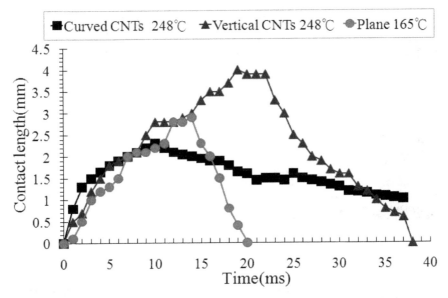

Figure 12. Contact line length versus time for droplets impinging on three surfaces at different temperatures

A digital pipette dispensed 0.1~2 μl of water onto CNTs nanotextured surface (hydrophilic surface) because the CNTs surfaces were hydrophilic modification. A 2.0 μl droplet of distilled water is placed on the heat surface at T_s=125 °C with three samples of oxidized silicon, vertical CNTs , or curved CNTs, respectively, and the water droplets were extended to a film as a function of very smaller CA was larger. The impact of a distilled water droplet upon a heated surface was investigated experimentally using a high-speed digital camera. Figure 13 illustrates images corresponding to the different characteristic stages of water droplet of 2.0μl evaporation on the oxidized silicon surface for a typical boiling condition. The time of initial solid-liquid contact, i.e. the first significant frame of a series, was taken to be the origin of the time axis (t_0 = 0). Thus, the last frame with the droplet still being visible corresponds to the droplet evaporation time. The droplet immediately boils when it contacts the surface, large bubble can be divided into film evaporation and nucleate boiling consequently dry-out are shown in figure 14. The curved CNTs surface enhanced capillary pumping effect and retained water during higher surface temperatures, evaporation rates are high and consequently the droplet lifetimes are short as shown in figure 15.

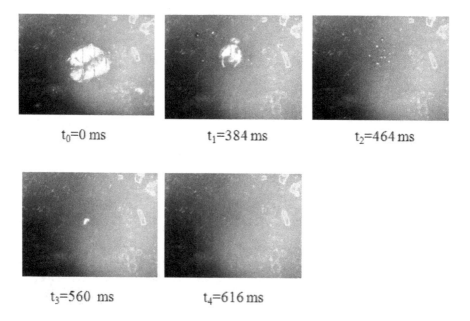

t_0=0 ms t_1=384 ms t_2=464 ms

t_3=560 ms t_4=616 ms

Figure 13. Images of top view corresponding to the characteristic stages of the droplet evaporation process for 2 μL water droplet impinging on plane oxidized silicon, typical boiling condition, Ts = 125 °C.

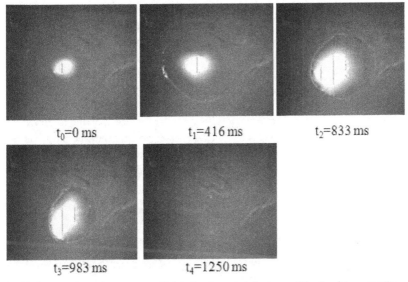

Figure 14. Images of top view corresponding to the characteristic stages of the droplet evaporation process for 2 μL water droplet impinging on the vertical CNTs surface, typical boiling condition, Ts = 125 °C.

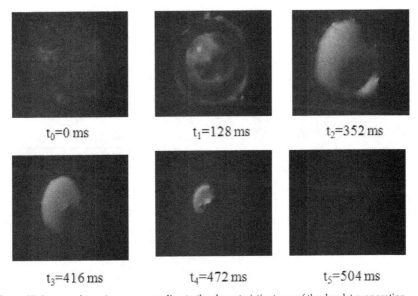

Figure 15. Images of top view corresponding to the characteristic stages of the droplet evaporation process for 2 μL water droplet impinging on the curved CNTs surface, typical boiling condition, Ts = 125 °C.

Figure 16 show the heat transfer curves of water spray cooling on the oxidized silicon and nanotextured surfaces when the cooling liquid is sprayed at a constant flow rate (11.7 ml/min) and standard ambient pressure (1 atm). In the evaporative zone, the heat flux range of 30~40 W/cm², the mean heat transfer coefficient of the curved CNTs surface was approximately 140% higher than that of the plane surface. Figure 17 shows the heat transfer coefficient with heat flux. The heat transfer coefficient derived from Eq. 1. The heat transfer coefficient of the spray cooling curve has a peak in the evaporative zone, and then, near 15 W/cm², it rapidly decreases. The maximum heat transfer coefficient occurs at curved CNTs surface because the effect of the evaporative latent heat increases to enhance heat transfer. In addition, the heat transfer coefficients of the CNTs nanotextured surface are in the range of 10~25 W/cm², which is higher than that of the plain surface due to the greater level of liquid wetting in the evaporative zone by spray cooling.

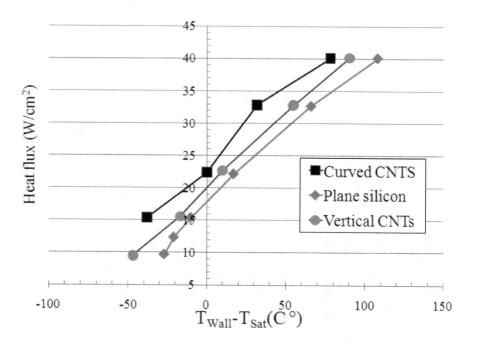

Figure 16. Comparison of heat flux between the oxidized silicon and nano-textured surfaces for spray cooling.

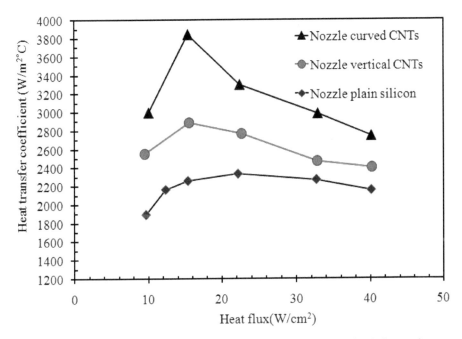

Figure 17. Heat transfer coefficients with variations in heat flux between the oxidized silicon and nano-textured surfaces for spray cooling.

5. Conclusions

The present study demonstrates the hydrodynamic characteristics of droplet impinging on heated surfaces using high-speed imaging and evaluates the heat transfer performance of surface temperature for water on both plane and nanotextured surfaces. Use of the nanotextured surface have resulted in a more uniform temperature profile at and near the impact area exhibiting lower minimum wall temperature especially at higher heat flux values. Nanotextured surfaces also yield lower static contact angle and enhanced film dynamics resulting in a noticeable enhanced heat transfer behavior. Results show that the curved CNTs surface enhanced capillary pumping effect and retained water during higher surface temperatures. This increased spreading over the nanotextured surface suggests a greater potential for increased liquid transport. In the evaporative zone, the heat flux range of 30~40 W/cm², the mean heat transfer coefficient of the curved CNTs surface was approximately 140% higher than of the oxidized silicon surface. This study offers insight into the design of nanotextured surface and new opportunities to achieve capillary pumping capability for microfluidic- based thermal management systems.

Nomenclature

C_p	heat capacity (J/kg•k)
d	droplet diameter (mm)
d_o	droplet diameter prior to impact (mm)
d^*	non-dimensional droplet diameter, d/d_o
G	groove width (mm)
g	gravitational constant (N/s^2)
I	Ampere (A)
H	droplet height (mm)
H^*	non-dimensional droplet height, H/d_o
t	time (s)
t^*	non-dimensional time, $t/(d_o/V_o)$
t_{ev}	droplet evaporation time (s)
T	temperature (°C)
T_r	rebounding temperature(°C)
T_s	surface temperature(°C)
V	voltage (V)
V_o	droplet impact velocity (m/s)

Greek symbols

ϱ_f	fluid density (kg/m^3)
μ	dynamic viscosity (Pa•s)
γ	surface tension (N/m)
θ	contact angle

Subscript

a	ambient
l	liquid phase
s	surface
v	vapor phase
sat	saturation
sub	subcooling
CHF	critical heat flux

Author details

Cheng Lin
The Second Research Division, Chung-Shan Institute of Science & Technology, Lung-Tan, Taoyan County, Taiwan, R.O.C

Acknowledgement

I would like to thank the National Science Council of the Republic of China, Taiwan, for financially supporting this research under Contract No. NSC 99-2221-E-007-028-MY2. The author also wishes to thank the NTHU ESS MEMS Laboratory for use of their facilities.

6. References

[1] Bar-Cohen A., Arik, M., and Ohad, M.," Direct liquid cooling of high flux micro and nano electronic components," Proceedings of the IEEE, Vol. 94, No. 8, pp. 1549-1570, 2006.

[2] Berber, S., Kwon, Y.-K., and Tománek, D., "Unusually high thermal conductivity of nanotubes," Phys. Rev. Lett., 84, pp. 4613-4616, 2000.

[3] Silk, E. A., Kim, J. G., Kiger, K. T., "Spray cooling of enhanced surfaces: Impact of structured surface geometry and spray axis inclination", Int. J. Heat Mass Transfer, Vol. 49, pp. 4910-4920, 2006.

[4] Coursey, J. S., Kim, J. G., Kiger, K. T. "Spray cooling of high aspect ratio open microchannels", J. Heat Transfer, Vol. 129, no. 8, pp.1052-1059, 2007.

[5] Pal, A., Joshi, Y.," Boiling of water at sub-atmospheric conditions with enhanced structures: Effect of liquid fill volume", J. Electron. Packag., Vol. 130, no. 1, 011010, 2008.

[6] Ahn, H. S., Sinha, N., Zhang, M., Banerjee, D., Fang, S. K., Baughman, R. H., " Pool boiling experiments on multiwalled carbon nanotube (MWCNT) forests", J. Heat Transfer, Vol. 128, pp. 1335-1342, 2006.

[7] Sodtke C. , Stephan P. ,"Spray cooling on micro-structured surfaces," Int. J. Heat Mass Transfer, Vol. 50, pp. 4089–4097, 2007.

[8] Amon, C.H., Yao, S. C., Wu, C. F. Wu, and Hsieh, C. C., " Microelectro- mechanical system-based evaporative thermal management of high heat flux electronics," ASME Journal of Heat Transfer, Vol. 127, pp. 66-75, 2005.

[9] Hsieh, C. C., Yao, S. C.," Evaporative heat transfer characteristics of a water spray on micro-structured silicon surfaces," Int. J. Heat Mass Transfer, Vol. 49, pp. 962–974, 2006.

[10] Visaria, M. and Mudawar, I.," Application of two-phase spray cooling for thermal management of electronic devices," IEEE Trans. Compon. Packag Technol., vol. 32, no. 4, pp.784–793, 2009.

[11] Ujereh, S., Fisher, T., Mudawar, I.," Effects of carbon nanotube arrays on nucleate pool boiling," Int. J. Heat Mass Transfer, Vol. 50, pp. 4023–4038, 2007.

[12] Li, C., Wang, Z., Wang, P. I., Peles, Y., Koratkar, N., and Peterson, G. P.," Nanostructured copper interfaces for enhanced boiling," Small, Vol. 4, No. 8, pp.1084–1088, 2008.

[13] Chen, R., Lu, M. C., Srinivasan, V., Wang, Z., Cho, H. H. Cho, and Majumdar, A." Nanowires for enhanced boiling heat transfer," Nano Lett., Vol. 9, No. 2, pp. 548-553, 2009.

[14] Khoo, H. S., Tseng, F. G., "Spontaneous high-speed transport of subnanoliter water droplet on gradient nanotextured surfaces", Appl. Phys. Lett., Vol. 95, 063108, 2009.

[15] Chen, M. H., Hsu, T. H., Chuang, Y. J., Tseng, F. G., " Dual hierarchical biomimic superhydrophobic surface with three energy states", Appl. Phys. Lett., Vol. 95, 023702, 2009.

[16] Lin, C., Chen, C. J., Chieng, C. C., Tseng, F. G., "Dynamic effects of droplet impingement on nanotextured surface for enhanced spray cooling ," The 16th International Conference on Solid-State Sensors, Actuators and Microsystems (Transducers'11), Beijing, China, June5-9, pp.1252-1255,2011.

[17] Carey, V. P., "Liquid-vapor phase-change phenomena", Hemisphere: Washington, DC, 1992.

[18] Deng, T., Varanasi, K. K., Hsu, M., Bhate, N., Keimel, C., Stein, J., Blohm, M., "Nonwetting of impinging droplets on textured surfaces", Appl. Phys. Lett., Vol. 94, 133109, 2009.

[19] Mudawar I. and Estes, K. A., "Optimizing and predicting CHF in spray cooling of a square surface," J. Heat Transfer, Vol. 118, pp.672-679, 1996.

[20] Lin, C., Tseng, F. G., Kan, H. C., Chieng, C. C. , " Numerical studies on micropart self-alignment using surface tension forces," J. Microfluidic and Nanofluidic, Vol. 6, pp.63-75, 2009.

Chapter 10

Condensate Drop Movement by Surface Temperature Gradient on Heat Transfer Surface in Marangoni Dropwise Condensation

Yoshio Utaka and Zhihao Chen

Additional information is available at the end of the chapter

1. Introduction

Marangoni dropwise condensation occurs in the condensation of a binary vapor mixture of a positive system, in which the surface tension of the mixture has a negative gradient with the mass fraction of the more volatile component, such as water–ethanol and water–ammonium mixtures. Thick condensate areas have higher liquid surface temperatures than thin areas; therefore, the surface tension flow is induced toward the peak of the condensate from thinner areas as a result of the vapor–liquid equilibrium and the variation in the surface tension in the binary vapor condensation of a positive system. This phenomenon differs essentially from so-called dropwise condensation on a hydrophobic surface, because there is a continuous thin liquid film between condensate drops and condensation occurs on a hydrophilic surface. This phenomenon was first reported by Mirkovich and Missen [1] in 1961 for a binary mixture of organic vapors. Ford and Missen [2] demonstrated that the criterion for instability of a condensate liquid film is $d\sigma/db > 0$, where b denotes the condensate film thickness. Fujii et al. [3] conducted an experimental investigation of the condensation of water–ethanol mixtures on a horizontal tube and observed several different condensation modes dependent on the concentration. Morrison and Deans measured the heat transfer characteristics of a water–ammonium vapor mixture and found that it exhibited enhanced heat transfer [4].

In recent years, Utaka and co-workers conducted research [5–9] on the dominant factors (surface subcooling, vapor mass fraction, and vapor velocity) in determining the condensation modes and heat transfer characteristics of Marangoni condensation. The major results on the heat transfer characteristics of Marangoni condensation were summarized in reference 9. Heat transfer was significantly enhanced for a low mass fraction of ethanol in a water–ethanol mixture. Murase et al. [10] studied Marangoni condensation of steam–ethanol mixtures using a horizontal condenser tube, and the results exhibited similar trends to those obtained by Utaka and Wang [7] for vertical surfaces.

The mechanisms of Marangoni condensation have also been studied. Hijikata et al. [11] presented a theoretical drop growth mechanism for Marangoni dropwise condensation. That is, the Marangoni effect occurs due to the surface tension difference which plays a more important role than the surface tension. Akiyama et al. [12] performed a 2-dimensional numerical simulation of the condensation of water-ethanol vapor on a horizontal heat transfer surface and found a 2 K temperature difference between the condensate film area and the crest of the condensate drop. Marangoni flow occurs in a condensate liquid and is driven by the surface tension gradient induced by the temperature difference. Utaka et al. [13] investigated the effect of the initial drop distance, which is the average distance between initially formed drops grown from a thin flat condensate film that forms immediately after a drop departs. They clarified that the initial drop distance is closely related to the heat transfer characteristics of Marangoni condensation. Furthermore, Utaka and Nishikawa [14] measured the thickness of condensate films on the tracks of departing drops and between drops using the laser extinction method, in which the proportion of laser light absorbed by the condensate liquid is dependent on the liquid thickness. The condensate film was approximately 1 μm thick and was strongly dependent on the initial drop distance and the heat transfer characteristics.

Marangoni condensation occurs due to the instability of Marangoni force acting on the condensate film. Condensate drops move spontaneously without any external forces when a bulk temperature gradient is applied to a horizontal heat transfer surface, only due to the imbalance of the surface tension distribution around the drops. This kind of phenomena could also occur in a low-gravity environment. This implies that condensate drops can be moved by applying a bulk surface temperature to a heat transfer surface. It is thus possible to remove a thick liquid film and large condensate drops by exploiting this spontaneous movement of condensate drops. A highly efficient heat exchanger could then be realized. Moreover, since non-uniform temperature distributions are often generated in heat exchangers, it is essential to clarify the heat transfer and condensate movement characteristics in Marangoni condensation when there is a temperature distribution on the heat transfer surface. It is also considered that the circulation of condensate driven by surface tension flow could be utilized in some heat transfer devices (e.g., a wickless heat pipe). Utaka and Kamiyama [15] examined the effect of the bulk surface tension gradient on condensate drop movement when a steady bulk temperature gradient was applied to horizontal and inclined heat transfer surfaces in the condensation of a water–ethanol vapor mixture. The condensate drops moved from the low-temperature side to the high-temperature side. The drop velocity increased with the surface tension gradient on the condensing surface and was independent of the drop size. Chen and Utaka [16] investigated the mechanisms and characteristics of drop movement on a horizontal condensing surface with a bulk temperature gradient for Marangoni dropwise condensation of a water–ethanol vapor mixture. In particular, experimental observations and measurements on the dominant factors affecting condensate drop movement were conducted, such as 1) bulk surface tension gradient, and 2) initial drop distance (adopted as a parameter for the Marangoni force and the condensate drop shape). The velocity of condensate drop movement was determined to correlate well with both the surface tension gradient and the initial drop distance.

In this chapter, the characteristics and mechanisms of condensate drop movement driven by a surface tension gradient in Marangoni dropwise condensation are summarized on the basis of the presented researches.

2. Marangoni dropwise condensation

2.1. Heat transfer characteristics in Marangoni dropwise condensation

In the condensation of a binary vapor mixture, such as water–ethanol vapor, the Marangoni force (indicated by the arrows in Fig. 1) pulls the condensate liquid from the periphery toward the peak along the surface of a condensate drop, whereby dropwise condensation occurs. The Marangoni force here is the driving force for condensate flow, which is considered to be caused by a surface tension difference on the condensate surface, based on the vapor–liquid equilibrium and the variation in the surface tension for a water–ethanol liquid mixture (see Fig. 2). This kind of phenomenon is referred to as 'Marangoni dropwise condensation'.

Utaka and Terachi [6] and Utaka and Wang [7] reported that significantly enhanced heat transfer could be realized by decreasing the thermal resistance of the condensate liquid in the Marangoni dropwise condensation of a water–ethanol vapor mixture. Utaka and Terachi [6] measured the condensation characteristics and clarified that surface subcooling is one of the dominant factors that determines the condensate and heat transfer characteristics of Marangoni condensation. More accurate measurements of a wider range of ethanol mass fraction and surface subcooling were conducted by Utaka and Wang [7], some of the results of which are shown in Fig. 3 and Fig. 4. Figure 3 shows the heat transfer coefficient of Marangoni condensation for water–ethanol vapor mixtures with various ethanol mass fractions. Figure 4 shows the variation in the ratio of the peak heat transfer coefficient of the mixture vapor to that of pure steam. These two figures indicate that the condensation heat transfer is significantly enhanced by the addition of an extremely small amount of ethanol and the heat transfer coefficient of the vapor mixture is approximately 8 times higher than that of pure steam.

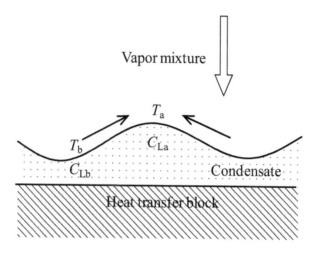

Figure 1. Mechanism for Marangoni dropwise condensation

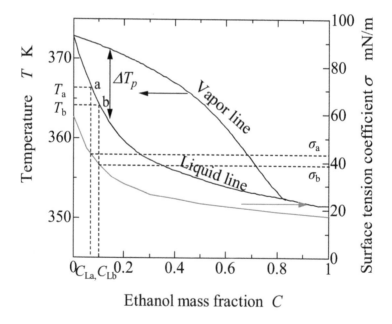

Figure 2. Vapor–liquid equilibrium and variation of surface tension coefficient for water–ethanol mixture

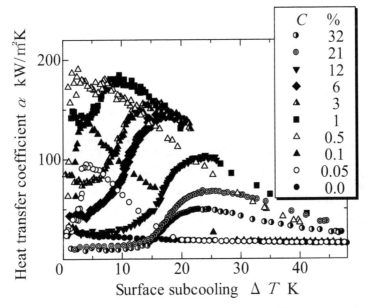

Figure 3. Condensation characteristics of water–ethanol vapor mixtures with various ethanol mass
fractions

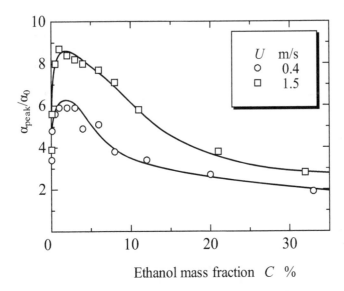

Figure 4. Variation in the ratio of the peak heat transfer coefficient of the vapor mixture to that of pure
steam

2.2. Relations among initial drop distance, Marangoni force and shape (angle) of condensate drop

In Marangoni dropwise condensation, small condensate drops initially form from a smooth and thin liquid film adjacent to the periphery of a large condensate drop after its departure. Figure 5 shows time-series microscopic images of the formation of small condensate drops. These initially formed drops are called initial drops and the average distance between the centers of the initial drops is defined as the initial drop distance.

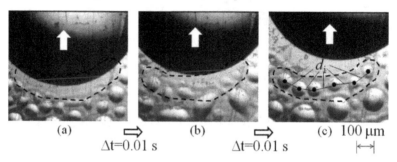

Figure 5. Initial drop formation process with initial drop distance (d_i) indicated

Condensate drops form in Marangoni dropwise condensation due to the Marangoni force acting on the surface of the condensate liquid. Therefore, the Marangoni force is considered to be closely related to the heat transfer mechanisms of Marangoni dropwise condensation. It is also reasonable that the formation of initial drops and the initial drop distance is determined by the strength of the Marangoni force. Thus, a close correlation exists between the initial drop distance and Marangoni force. Consequently, the initial drop distance is adopted as an important parameter of the heat transfer characteristics and mechanisms of Marangoni condensation in the studies of Utaka et al. [13] and Utaka and Nishikawa [14]. Figures 6(a) and (b) show respective plots of the heat transfer coefficient and the initial drop distance as a function of surface subcooling based on data measured by Utaka et al. [13]. The initial drop distances have U-shaped curves with minima that correspond to distances in the range 30–150 μm, depending on the surface subcooling and the mass fraction of ethanol. Surface subcooling at the minimum initial drop distances coincides with that at the maximum heat transfer coefficient. Utaka and Nishikawa [14] investigated the relationship between the liquid film thickness and the initial drop distance (Fig. 7) for a water–ethanol mixture using the laser extinction method. A condensate liquid film of approximately 1 μm thickness remained after sweeping by departing drops and between condensate drops. The minimum condensate film thickness decreased with initial drop distance for surface subcooling lower than the maximum heat transfer point, even when the condensation rate increased.

These two studies demonstrated that there is a close relationship between the heat transfer coefficient and characteristic parameters such as the initial drop distance and the minimum condensate thickness. In the surface subcooling region near the maximum heat transfer coefficient, the initial drop distance and minimum film thickness tend to assume minimum values as a result of the driving force being a maximum, due to the surface tension gradient on the condensate surface. Thus, when the initial drop distance decreases, heat transfer is enhanced by thinning of the condensate film that could result in a reduction in the thermal resistance of the condensate. In addition, the condensate drop shape changes with increasing Marangoni force and the condensate film becomes thinner, even when condensation rate increases. This implies that the drop height increases as the drops approach hemispherical shapes due to an increase in the Marangoni force. The correlation among the Marangoni force, initial drop distance and shape (angle) of condensate drops, as

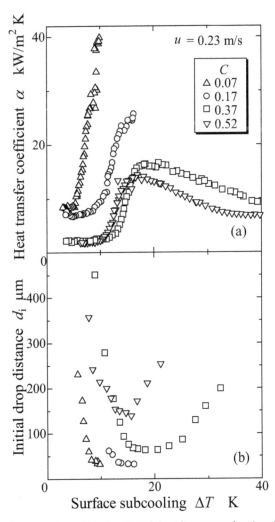

Figure 6. Variation of heat transfer coefficient and initial drop distance as a function of surface subcooling

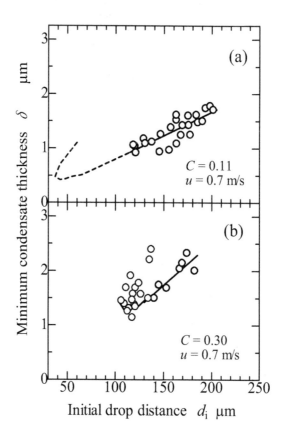

Figure 7. Variation of minimum condensate film thickness as a function of initial drop distance

shown in Fig. 8, could be inferred on the basis of these experimental results. For certain   mass fractions of ethanol, the experimental condition of surface subcooling determines the strength of the Marangoni force, and thus the initial drop distance and shape of the condensate drop are also determined. Therefore, if any one of the three factors is known, the two other factors can be determined based on the corresponding correlations. The qualitative correlations are inferred from the experimental results. The quantitative correlations were experimentally studied and are introduced in the following section.

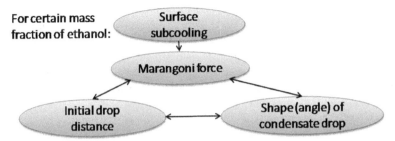

Figure 8. Correlation between Marangoni force, initial drop distance and condensate drop shape

3. Spontaneous movement of condensate drops in Marangoni dropwise condensation

When a bulk temperature gradient is applied to a horizontal heat transfer surface or in a low-gravity environment under a Marangoni condensation field, condensate drops move spontaneously without external forces. The reason for condensate drop movement is considered to be as follows. A Marangoni force (F_H or F_L in Fig. 9, the letter 'H' represents the high temperature side and 'L' represents the high temperature side.) is induced by the difference in surface tension on the condensate surface in Marangoni dropwise condensation. The condensate near the periphery of a condensate drop is pulled along the condensate liquid surface toward the peak of the drop. A reactive force against the surface tension flow caused by the Marangoni force is induced at the drop periphery. When there is no bulk temperature gradient, the reactive force is uniform around the drop periphery and averages out over time, so that the drop does not move to a large extent. In contrast, when a bulk temperature gradient is applied to a horizontal heat transfer surface, the horizontal component of the reactive force (F_{HX} or F_{LX} in Fig. 9, the letter 'X' represents the horizontal component.) around a condensate drop becomes nonuniform, as shown in Fig. 9. Consequently, condensate drops move spontaneously on the heat transfer surface without external forces.

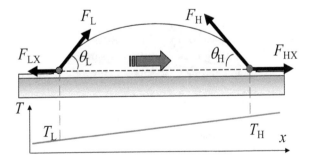

Figure 9. Schematic diagram of the driving force for condensate drop movement

It is considered that the imbalance of the reactive force is determined by the bulk surface tension gradient of the condensate liquid. Consequently, the velocity of condensate drop movement is considered to be affected by the bulk surface tension gradient. The bulk surface tension gradient is calculated from the surface tension difference, which corresponds to the time-averaged surface temperature distribution of the extremely thin liquid film covering the heat transfer surface. The horizontal component of the Marangoni force depends on the overall magnitude of the Marangoni force and the shape (angle) of the condensate drop. Therefore, it is conceivable that the movement of the condensate drop is also determined by these two factors. Based on the correlation between the Marangoni force, initial drop distance and the angle of the condensate drop shown in Fig. 8, it follows that condensate drop movement is also affected by these three factors. Utaka and co-workers [15, 16] have focused on these factors and carried out several experimental and numerical studies on the characteristics and mechanisms of condensate drop movement in Marangoni dropwise condensation.

3.1. Experimental apparatus

Figure 10 shows a schematic of a typical experimental system. A vapor mixture is generated by electrically heating a water–ethanol mixture with a certain mass fraction in a vapor generator. The vapor is partially condensed on a heat transfer block, and is almost completely condensed in an auxiliary condenser after passing through the condensing chamber. The vapor pressure is maintained close to atmospheric pressure by a small opening to the atmosphere between the auxiliary condenser and condensate receiver. The condensate is fed back into the vapor generator after deaeration to remove non-condensable gases dissolved in the condensate. In addition, the vapor is made to flow in the opposite direction to the condensate drop movement to distinguish the driving force of drop movement from the shear force of the vapor flow. Figure 11 shows a schematic of the condensing chamber, where the condensate drop behavior is observed through front and side windows.

Figure 12 shows a schematic of the heat transfer block, which was made of brass with a surface area of 20×20 mm² that was positioned horizontally for the experiments. A triangular cross-section of constantan, which has low thermal conductivity, was soldered onto the cooling surface of the heat transfer block. This allowed a bulk temperature gradient to be applied to the heat transfer surface by uniformly cooling the constantan surface with multiple water jet spray. Temperature was measured using thermocouples located inside the heat transfer block, and the surface temperature distribution was determined by two-dimensional extrapolation. The heat transfer surface was coated with titanium dioxide to make it hydrophilic to distinguish it from dropwise condensation on a hydrophobic surface. Experiments were conducted continuously using quasi-steady-state measurements, in which the temperature of the cooling water was changed very slowly.

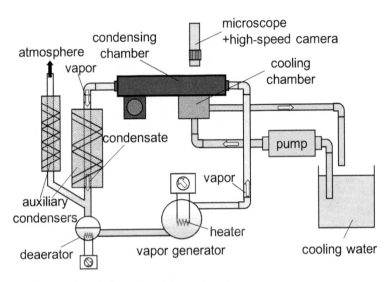

Figure 10. Schematic of a typical experimental apparatus setup

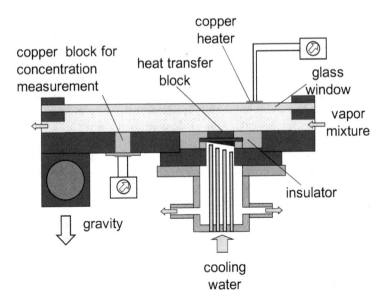

Figure 11. Schematic of the condensing chamber

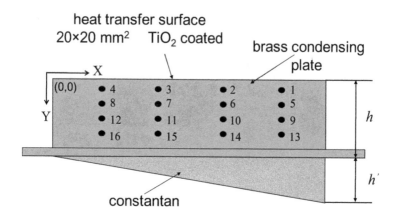

Figure 12. Schematic of the heat transfer block

3.2. Variations of condensate drop shape, initial drop distance and heat transfer coefficient against surface subcooling

To confirm the correlations among the Marangoni force, initial drop distance and shape of condensate inferred in section 2.2, experimental studies were conducted to investigate the quantitative relations. The angle of the condensate drop, initial drop distance and heat transfer coefficient were experimentally measured and the relations are discussed. A profile image of a condensate drop taken through the side view window is shown in Fig. 13. Vapor flows from the right side to the left side, which is the high-temperature side of the heat transfer surface, as does the condensate drop. The angle between the surface of the condensate drop and the heat transfer surface near the drop base as shown in the profile image is defined as the angle of the condensate drop. The angle in the direction of forward movement is the advancing angle θ_a, and that at the opposite side is the receding angle, θ_r. Since similar tendencies in the variation of advancing and receding angles were observed, an average value of the advancing and receding angle for single condensate drops was calculated. The variation of average condensate drop angle, initial drop distance and the heat transfer coefficient as a function of surface subcooling are shown in Fig. 14.

Several tendencies are evident in Fig. 14. For each mass fraction of ethanol, as with the previous results, the heat transfer coefficient increases and the initial drop distance decreases with increasing surface subcooling. In addition, the average condensate drop angle increases with increasing surface subcooling. This indicates that the decrease in the initial drop distance corresponds to an increase in the angle of the condensate drop. The maximum average angle of a condensate drop was approximately 35-45°, which is slightly smaller than the typical contact angle of a condensate drop on a hydrophobic surface. Moreover, for the same surface subcooling, a higher heat transfer coefficient and smaller initial drop distance were realized for a smaller mass fraction of ethanol.

The experimental results indicate that greater surface subcooling or lower mass fraction of ethanol gives a smaller initial drop distance, and the average angle of the condensate drop is larger due to the stronger Marangoni force. Therefore, it was confirmed that the three main factors have quantitative correlations. In addition, there was a large amount of scatter in the data for the condensate drop angle, which is caused by frequent coalescence when the drops are moving or by variation of the temperature distribution on the heat transfer surface. This scatter is considered to be an essential characteristic of Marangoni dropwise condensation. Therefore, in the three important factors, the Marangoni force cannot be measured and the condensate drop angle has large amount of scatter. In contrast with the other two factors, initial drops grow from a thin flat condensate film that appears immediately after a drop departs, the state before the initial drops form is relatively stable, and thus, the measurement of the initial drop distance has good repeatability. Therefore, it is appropriate to adopt the initial drop distance as the dominant parameter of Marangoni dropwise condensation that represents the Marangoni force and the shape of a condensate drop.

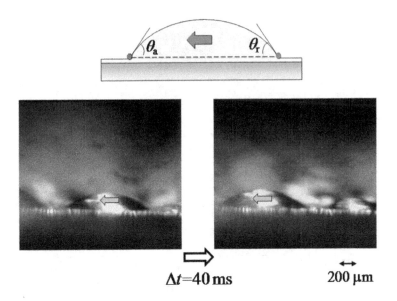

$\Delta t = 40\,\text{ms}$ $200\,\mu\text{m}$

Figure 13. Profile image of condensate drop shape

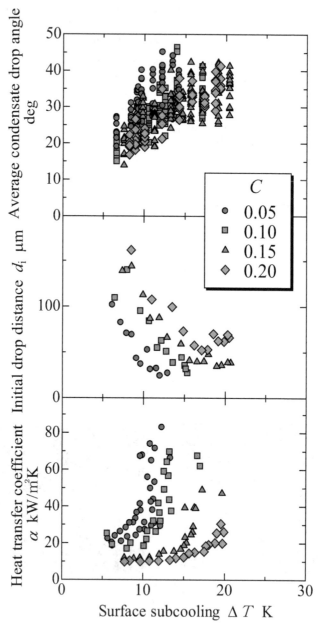

Figure 14. Variation of condensate drop angle, initial drop distance and heat transfer coefficient as a function of surface subcooling for various mass fractions of ethanol

3.3. Effect of bulk surface tension gradient on velocity of drop movement

Utaka and Kamiyama [15] examined the effect of the bulk surface tension gradient on condensate drop movement when a steady bulk temperature gradient was applied to horizontal and inclined heat transfer surfaces during the condensation of a water–ethanol vapor mixture. Figure 15 shows images of condensate drop movement on the horizontal heat transfer surface. The upper part of the image is the high-temperature side and the lower part is the low-temperature side. The condensate drops move from the low-temperature side to the high-temperature side. The variations of condensate drop velocity are shown for various ethanol mass fractions as a function of the bulk surface tension gradient in Fig. 16. The drop velocity increased with increasing surface tension gradient on the condensing surface and was independent of the drop size. Moreover, although there is a large scatter in the drop velocities due to frequent coalescence of the condensate drops, qualitatively similar tendencies of drop velocity were shown.

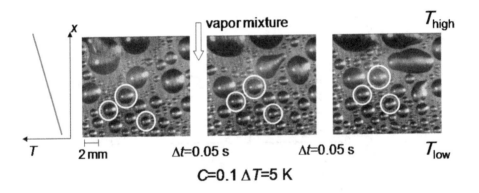

Figure 15. Appearance of condensate drop movement

3.4. Effect of initial drop distance on velocity of drop movement

Chen and Utaka [16] investigated the affects of the Marangoni force and the condensate drop angle on the velocity of condensate drop movement. As discussed in section 3.2, the initial drop distance was adopted as the dominant parameter representing the Marangoni force and angle of condensate drop. The variations of velocity of drop movement as functions of the initial drop distance and bulk surface tension gradient are shown in Figs. 17 and 18, respectively. The drop velocities vary significantly, so that an average velocity of all condensate drops was adopted for each set of conditions.

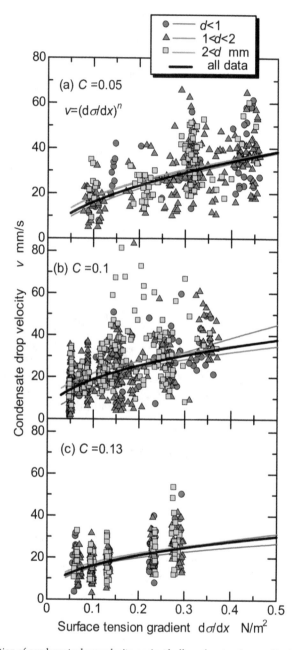

Figure 16. Variation of condensate drop velocity against bulk surface tension gradient

Figures 17 show plots of the average drop velocity as a function of the initial drop distance for ethanol mass fractions C, of 0.1, 0.15, 0.2, and 0.25 for six different bulk surface tension gradients. All experiments were performed in the surface subcooling range for which the heat transfer coefficient is less than its maximum value. Distinct trends between the average drop velocity and the initial drop distance are observed in Fig. 17. Qualitatively similar tendencies are evident and the curves have maximum values for each ethanol mass fraction and each surface tension gradient. When the initial drop distance decreases, the average drop velocity initially increases and then decreases after reaching a maximum value at almost the same surface tension gradient for all ethanol mass fractions. For example, for $C =$ 0.15 and a bulk surface tension gradient of 0.2–0.3 N/m², the average drop velocity increases from $v = 16$ mm/s to a maximum value of $v = 21$ mm/s when the initial drop distance decreases from $d_i = 190$ to 130 μm. The average velocity then tended to decrease to $v = 0$ mm/s as d_i decreased to 25 μm. While the changes in drop velocity were gradual at relatively high bulk surface tension gradients, the drop velocities over the entire range of initial drop distances decreased significantly over a smaller range of bulk surface tension gradients for all ethanol mass fractions. Although it is not surprising that the driving force is approximately 0 at low surface tension gradients in the ranges –0.05 to 0 N/m² and 0–0.05 N/m² for all mass fractions, it is notable that the driving force is also very small for small d_i in the initial drop distance range of 30–40 μm, even at high surface tension gradients.

Figure 18 shows the effect of the bulk surface tension gradient on the drop velocity for four initial drop distances and for ethanol mass fractions C, of 0.1, 0.15, 0.2, and 0.25. Data with similar initial drop distances as those in Fig. 17 were selected and were plotted together and fitted with lines that pass through the origin. The drop velocity increases linearly with increasing bulk surface tension gradient for each initial drop distance range. Furthermore, the rate of increase in the drop velocity with the bulk surface tension gradient increases with increasing initial drop distance in the lower ranges of initial drop distance up to the peak average drop velocity shown in Fig. 17. Similar increasing rate were obtained in the larger ranges of initial drop distance for each mass fraction. The effect of the surface tension gradient on the drop velocity became stronger when the initial drop distance approached values that give rise to the maximum velocities shown in Fig. 17. For example, when the bulk surface tension gradient is –0.05 N/m², the velocity of condensate drops is around 0 mm/s when the initial drop distances are in the range of 30–35 μm. The velocity then increases with increasing bulk surface tension gradient; the velocity is 10 mm/s at a bulk surface tension gradient of 0.34 N/m². For larger initial drop distances, the increase in the drop velocity as a function of bulk surface tension gradient becomes more rapid when the initial drop distance was in the range of 60–80 μm. This increase becomes much more rapid and the velocity increases from 3.6 to 18.9 mm/s in the initial drop distance range of 155–255 μm, when the corresponding bulk surface tension gradient is increased from 0.09 to 0.28 N/m². For comparatively large initial drop distances in the ranges of 155–215 μm and 200–270 μm, the variations in the drop velocity as a function of the bulk surface tension gradient were similar, as shown in Fig. 18(b).

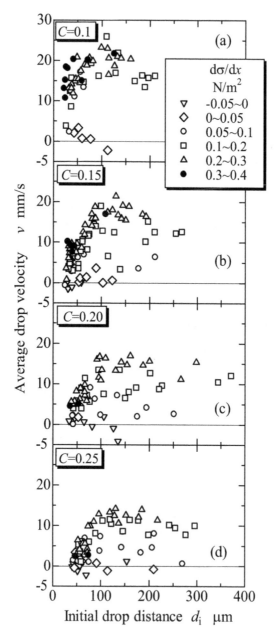

Figure 17. Variation of average drop velocity as a function of initial drop distance for various ethanol mass fractions and bulk surface tension gradients

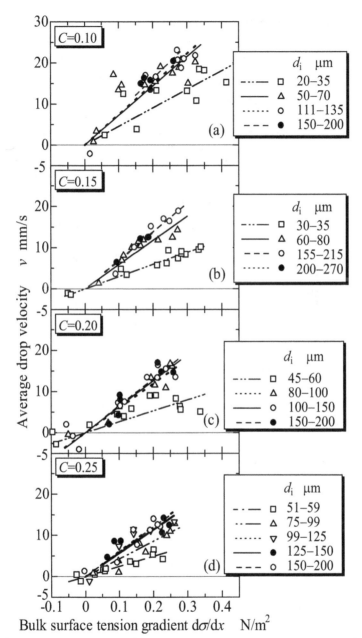

Figure 18. Variation of average drop velocity as a function of bulk surface tension gradient for various
ethanol mass fractions and initial drop distances

3.5. Mechanisms of condensate drop movement

Experimental studies on the characteristics (effects of several parameters) of drop movement under a bulk temperature gradient on a heat transfer surface have been conducted for Marangoni dropwise condensation of water-ethanol vapor. However, the essential factor relating to Marangoni dropwise condensation and the condensate drop movement, the Marangoni force, cannot be experimentally measured. Therefore, to better understand the relationship between the Marangoni force and condensate drop movement, numerical simulation of the spontaneous movement of condensate drops was conducted using the volume of fluid (VOF) method. In this section, the 3-dimensional phenomenon of condensate drop movement was simulated using a 2-dimensional calculation in the domain presented in Fig. 19; therefore, only qualitative discussion is presented in this section.

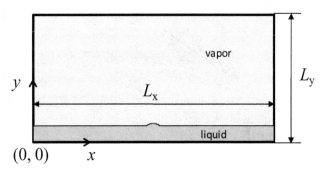

Figure 19. Calculation domain used for the numerical simulation

3.5.1. Governing equations

3.5.1.1. Liquid phase

In the calculation, the liquid phase was treated as incompressible. The continuity, momentum, and energy equations were solved.

$$\frac{\partial u_L}{\partial x} + \frac{\partial v_L}{\partial y} = 0 \tag{1}$$

$$\frac{\partial u_L}{\partial t} + u_L \frac{\partial u_L}{\partial x} + v_L \frac{\partial u_L}{\partial y} = g_x - \frac{1}{\rho_L} \frac{\partial P}{\partial x} + v_L \left(\frac{\partial^2 u_L}{\partial x^2} + \frac{\partial^2 u_L}{\partial y^2} \right) \tag{2}$$

$$\frac{\partial v_L}{\partial t} + u_L \frac{\partial v_L}{\partial x} + v_L \frac{\partial v_L}{\partial y} = g_y - \frac{1}{\rho_L} \frac{\partial P}{\partial y} + v_L \left(\frac{\partial^2 v_L}{\partial x^2} + \frac{\partial^2 v_L}{\partial y^2} \right) \tag{3}$$

$$\frac{\partial T_L}{\partial t} + u_L \frac{\partial T_L}{\partial x} + v_L \frac{\partial T_L}{\partial y} = \frac{\lambda_L}{\rho_L C_{P,L}} \left(\frac{\partial^2 T_L}{\partial x^2} + \frac{\partial^2 T_L}{\partial y^2} \right) \tag{4}$$

$$\frac{\partial C_L}{\partial t} + u_L \frac{\partial C_L}{\partial x} + v_L \frac{\partial C_L}{\partial y} = D_L \left(\frac{\partial^2 C_L}{\partial x^2} + \frac{\partial^2 C_L}{\partial y^2} \right) \tag{5}$$

$$\frac{\partial F}{\partial t} + \frac{\partial u_L F}{\partial x} + \frac{\partial v_L F}{\partial y} = 0. \tag{6}$$

3.5.1.2. Vapor phase

In the vapor phase, the vapor velocity in the y-direction was determined by the condensation rate at the vapor-liquid interface, and the vapor velocity in the x-direction was ignored. The diffusion equation was solved based on the assigned vapor velocity. After the results of diffusion equation were obtained, the temperature of the vapor mixture was calculated from the mass fraction based on the vapor line in the relation of vapor-liquid equilibrium.

$$\rho_V v_V = \dot{m} \tag{7}$$

$$\frac{\partial C_V}{\partial t} + v_V \frac{\partial C_V}{\partial y} = D_V \left(\frac{\partial^2 C_V}{\partial x^2} + \frac{\partial^2 C_V}{\partial y^2} \right) \tag{8}$$

$$T_V = f(C_V). \tag{9}$$

3.5.1.3. Vapor-liquid interface

The velocity distribution at the vapor-liquid interface is calculated by taking account of the effect of the surface tension gradient (stress balance at the vapor-liquid interface). This was used as the velocity boundary condition of the liquid phase. The temperature and mass fraction of ethanol at the interface were calculated based on the vapor-liquid equilibrium, energy balance and mass balance.

Condensation rate:

$$\rho_V v_V = \dot{m} \tag{10}$$

Energy balance:

$$-\lambda_L \frac{\partial T_L}{\partial y} = \Delta h_V \dot{m} \tag{11}$$

Mass balance:

$$\dot{m}_E = \rho_V D_V \frac{\partial C_V}{\partial y} + C_V \dot{m} \tag{12}$$

Mass fraction of liquid in the cell at the interface:

$$C_L = \frac{\dot{m}_E}{\dot{m}} \tag{13}$$

Vapor-liquid equilibrium:

$$C_V = f(T_{surf})$$ (14)

$$C_L = f(T_{surf})$$ (15)

Relationship between the surface tension coefficient and concentration of the liquid:

$$\sigma = f(C_L)$$ (16)

Increase of F caused by condensation:

$$\frac{\partial F}{\partial t} = \frac{1}{\rho_L}\frac{\dot{m}}{\Delta y}$$ (17)

Stress balance at the vapor-liquid interface:

$$\left(P_V - P_L + \frac{\sigma}{R}\right)n_i = \left(-\tau_{L,ij}\right)n_k + \frac{\partial \sigma}{\partial x_i}$$ (18)

3.5.2. Boundary and initial conditions

Considering the real phenomenon and the computation time, the calculation is conducted in a relatively small region of 600×200 µm² (Fig. 19). The boundary conditions are summarized in Table 1. The boundary at $y=0$ is set as the condensing surface, and $y=L_y$ is the steady temperature/concentration boundary. In addition, the boundary condition at $x=0$, L_x is the free inlet/outlet flow. A thin (1.5 µm) liquid film is set initially on the condensing surface. A tiny protuberance is also given in the center of the initial liquid film as a disturbance. During the calculation, a certain mass fraction of ethanol vapor and the corresponding vapor line temperature are assigned to the boundary of $y=L_y$, and the temperature gradient (the right side is set as the high-temperature side) was assigned directly to the condensing surface of $y=0$. During the calculation, a constant temperature was initially assigned to the condensing surface. After the temperature/concentration distribution in the calculation region became close to that for the actual phenomenon with elapse of time, the temperature gradient was applied to the condensing surface.

Furthermore, the basic equations were discretized using a staggered grid. The convective term was approximated by a 1st–order upwind difference and the diffusion term by 2nd–order central difference. Pressure was calculated implicitly. Other variables such as velocity, temperature and mass fraction were calculated explicitly. The velocity field in the calculation region was calculated using the SOLA method. In addition, the variations of F at the vapor-liquid interface were calculated using the donor-acceptor method.

$y = L_y$	$\partial u/\partial x = 0$, $v = 0$	$T = T_0$	$C = C_0$
$y = 0$	$u = 0$, $v = 0$	$T = T_w(x)$	$\partial C/\partial y = 0$
$x = 0$	$\partial u/\partial x = 0$, $\partial v/\partial y = 0$	$\partial T/\partial x = 0$	$\partial C/\partial x = 0$
$x = L_x$	$\partial u/\partial x = 0$, $\partial v/\partial y = 0$	$\partial T/\partial x = 0$	$\partial C/\partial x = 0$

Table 1. Summary of the boundary conditions employed for the numerical simulation

3.5.3. Calculation results and discussion

3.5.3.1. Variation of liquid film and shape of condensate drop

The calculation results for an ethanol mass fraction of $C = 0.09$ are shown in Figs. 20–23. The variation in the thickness of the condensate film and the form of the free surface are shown in Fig. 20. The condensate film became thicker over time. Several condensate drops formed and became larger on the condensing surface including the spot where the initial disturbance was located.

3.5.3.2. Angle of the condensate drop

Figure 21 shows a comparison of condensate drops forming on condensing surfaces with different surface subcooling (ΔT=6 and 10 K). Condensate drops with similar diameters were selected and the shape of the drops was compared. Developed condensate drops were investigated to avoid the influence of initial conditions. The condensate drop was higher for larger surface subcooling ($\Delta T = 10$ K). Thus, the angle of the condensate drop becomes larger when the surface subcooling is larger, which is in agreement with the experimental results and indicates the condensate drop angle becomes larger because of the stronger Marangoni force.

3.5.3.3. Driving force of drop movement

To investigate the driving force of condensate drop movement, the momentum of condensate liquid pulled into a condensate drop by the Marangoni force around the periphery was calculated. In the two-dimensional simulation, the momenta on the high and low-temperature sides of a condensate drop were calculated. The qualitative relation between the drop movement and the imbalance of momentum in the horizontal direction is discussed. In addition, the momentum was calculated at the position where the condensate film around the periphery of a condensate drop is the thinnest (in the valley around the base of a condensate drop).

The experimental results obtained so far indicate that there is a large amount of scatter in the velocities and angles of condensate drops, due to the coalescence of drops or unstable temperature distributions near the periphery of drops. Similar to the experimental results, it is considered that the calculation results also vary significantly around the average values in the numerical simulation. Thus, to avoid the influence of adjacent condensate drops, the condensate drop formed in the vicinity of the center of a condensing surface (Fig. 20) was selected. Because it was considered that the characteristics of a relatively isolated condensate drop in the numerical simulation is nearly equal to that of condensate drops in the experiments.

The aspects of growth and movement of a condensate drop after the temperature gradient was applied are shown in Fig. 22 for $C=0.09$ and $\Delta T =10$ K. The crests of the condensate drops are indicated by open circles. In addition, the horizontal component of momentum at the high/low-temperature side of the condensate drop and the temperature difference of the liquid surface between the high- (right side) and low- (left side) temperature sides of the condensing surface (boundary condition) are shown in Figs. 23(a) and (b), respectively. The surface temperature of the condensate on the high-temperature side of the condensate drop is higher than that of the low-temperature side, and the horizontal momentum of the condensate liquid is larger on the high-temperature side than that on the low-temperature side.

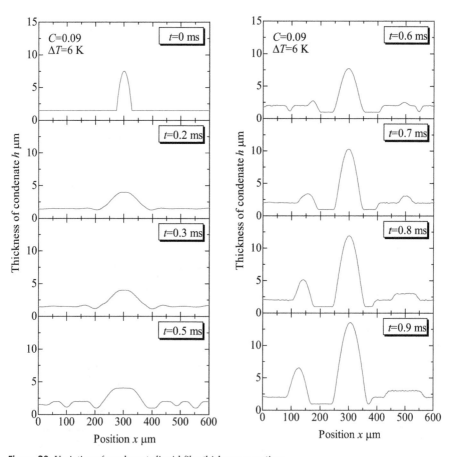

Figure 20. Variation of condensate liquid film thickness over time

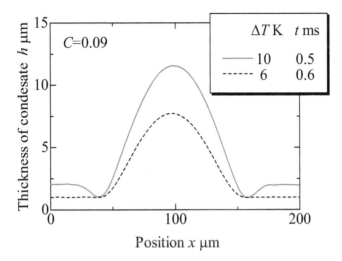

Figure 21. Comparison of condensate drop shape for different subcooling temperatures

These results correspond to those given in Fig. 22, where the growing condensate drop moves towards the high-temperature side. In conclusion, the condensate drop movement is in the direction of the side with the larger momentum of condensate liquid being pulled into the condensate drop by the Marangoni force. Thus, it could be inferred that an imbalance of the horizontal component of the Marangoni force is the driving force for condensate drop movement.

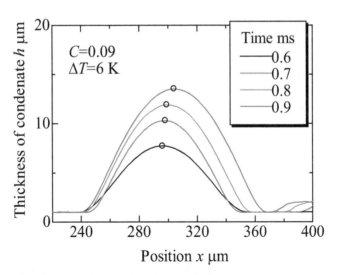

Figure 22. Growth and movement of a condensate drop over time

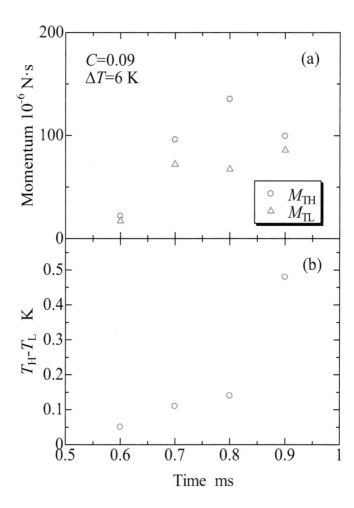

Figure 23. (a) Horizontal momentum of a condensate liquid driven into a condensate drop, and (b) surface temperature difference between the low- and high-temperature side of a condensate drop

4. Conclusion

Condensate drops move spontaneously on a heat transfer surface with a bulk temperature gradient in Marangoni dropwise condensation. It is conceivable that the velocity of a condensate drop is determined by the bulk surface tension gradient, Marangoni force, and the shape of the condensate drop. With a focus on these three factors, experimental and numerical studies were conducted on the spontaneous movement of condensate drops in the Marangoni condensation of a water–ethanol mixture. The results are summarized as follows:

1. Condensate drops move from the low-temperature to high-temperature side of a heat transfer surface. The velocity of condensate drop movement increases with the bulk surface tension gradient.
2. There are correlations among the Marangoni force, initial drop distance and angle of the condensate drop. It is appropriate to adopt the initial drop distance as a dominant parameter to express the characteristics of drop velocity.
3. When the initial drop distance decreases, the average drop velocity initially increases and then decreases after reaching a maximum at almost the same surface tension gradient. The average drop velocity increases linearly with bulk surface tension gradient for each initial drop distance range. The rate of increase in the drop velocity increases with the increasing initial drop distance.
4. Condensate drop movement is directed toward the side with a larger momentum of condensate liquid being pulled into the condensate drop by Marangoni force. It could be inferred that an imbalance of the horizontal component of Marangoni force is the driving force for condensate drop movement.

Nomenclature

C		Mass fraction of ethanol vapor
C_P	[J/kg·K]	Specific heat at constant pressure
D	[m²/s]	Diffusion coefficient
d	[mm]	Diameter of condensate drop
d_i	[μm]	Initial drop distance
F		VOF function, Force
g	[m/s²]	Gravity acceleration
Δh	[kJ/kg]	Latent heat
M	[N·s]	Momentum
$\dot{m}$	[kg/m²·s]	Mass flux of condensation
P	[kPa]	Pressure
T	[K]	Temperature

ΔT	[K]	Surface subcooling
t	[s]	Time
U	[m/s]	Velocity of vapor mixture
u	[m/s]	Horizontal velocity
v	[mm/s], [m/s]	Velocity of drop movement, Vertical velocity
x	[m]	Cartesian axis direction
y	[m]	Cartesian axis direction

Greek characters

α	[kW/m²k]	Heat transfer coefficient
θ	[°]	Angle of condensate drop
σ	[mN/m]	Surface tension coefficient
δ	[µm]	Minimum condensate thickness
λ	[W/m·K]	Thermal conductivity
υ	[m²/s]	Kinematic viscosity
ρ	[kg/m³]	Density
τ	[Pa]	Shear stress

Subscripts

a	Advancing angle
r	Receding angle
surf	Vapor-liquid interface
E	Ethanol
L	Liquid phase, low-temperature side
V	Vapor phase
H	High-temperature side
x	Horizontal

Author details

Yoshio Utaka and Zhihao Chen
Division of Systems Research, Faculty of Engineering, Yokohama National University, Japan

5. References

[1]   Mirkovich, V.V. and Missen, R.W. (1961). Non-Filmwise Condensation of Binary Vapor of Miscible Liquids, *Can. J. Chem. Eng.*, Vol. 39, pp. 86–87.

[2] Ford, J.D. and Missen, R.W., (1968). On the Conditions for Stability of Falling Films
     Subject to Surface Tension Disturbances; the Condensation of Binary Vapor, *Can. J.
     Chem. Eng.*, Vol. 48, pp. 309–312.
[3] Fujii, T., Osa, N. and Koyama, S., (1993). Free Convective Condensation of Binary Vapor
     Mixtures on a Smooth Horizontal Tube: Condensing Mode and Heat Transfer
     Coefficient of Condensate, *Proc. US Engineering Foundation Conference on Condensation
     and Condenser Design*, St. Augustine, Florida, ASME, pp. 171–182.
[4] Morrison, J.N.A. and Deans, J., (1997). Augmentation of Steam Condensation Heat
     Transfer by Addition of Ammonia, *Int. J. Heat Mass Transfer*, Vol. 40, pp. 765–772.
[5] Utaka, Y. and Terachi, N., (1995). Measurement of Condensation Characteristic Curves
     for Binary Mixture of Steam and Ethanol Vapor, *Heat Transfer-Japanese Research*, Vol. 24,
     pp. 57–67.
[6] Utaka, Y. and Terachi, N., (1995). Study on Condensation Heat Transfer for
     Steam–Ethanol Vapor Mixture (Relation between Condensation Characteristic Curve
     and Modes of Condensate), *Trans. Jpn. Soc. Mech. Eng.*, Series B, Vol. 61, No. 588, pp.
     3059–3065.
[7] Utaka, Y. and Wang, S., (2004). Characteristic Curves and the Promotion Effect of
     Ethanol Addition on Steam Condensation Heat Transfer, *Int. J. Heat Mass Transfer*, Vol.
     47, pp. 4507–4516.
[8] Utaka, Y. and Kobayashi, H., (2003). Effect of Vapor Velocity on Condensation Heat
     Transfer for Water–Ethanol Binary Vapor Mixture, *Proceedings of 6th ASME-JSME
     Thermal Engineering Conference*.
[9] Utaka, Y., (2011). Marangoni Condensation Heat Transfer. In: Belmiloudi, A., editor.
     Experimental Investigations and Industrial Systems, Heat Transfer-Theoretical
     Analysis, InTech. ISBN 978-953-307-226-5. pp. 327-350.
[10] Murase, T., Wang, H.S. and Rose, J.W., (2007). Marangoni condensation of steam-
     ethanol mixtures on a horizontal tube, *Int. J. Heat Mass Transfer*, Vol. 50, pp. 3774–3779.
[11] Hijikata, K., Fukasaku, Y. and Nakabeppu, O., (1996). Theoretical and Experimental
     Studies on the Pseudo-Dropwise Condensation of a Binary Vapor Mixture, *J. Heat
     Transfer*, Vol. 118, pp. 140–147.
[12] Akiyama, H., Nagasaki, T. and Ito, Y., (2001). A Study on the Mechanism of Dropwise
     Condensation in Water-Ethanol Vapor Mixture, *Thermal Science & Engineering*, Vol. 9,
     No. 6, pp. 19-27.
[13] Utaka, Y., Kenmotsu, T. and Yokoyama, S., (1998). Study on Marangoni Condensation
     (Measurement and Observation for Water and Ethanol Vapor Mixture), *Proceedings of
     11th International Heat Transfer Conference*, Vol. 6, pp. 397–402.
[14] Utaka, Y. and Nishikawa, T., (2003). Measurement of Condensate Film Thickness for
     Solutal Marangoni Condensation Applying Laser Extinction Method, *J. Enhanc. Heat
     Transf.*, Vol. 10, No. 1, pp. 119–129.

[15] Utaka, Y. and Kamiyama, T., (2008). Condensate Drop Movement in Marangoni Condensation by Applying Bulk Temperature Gradient on Heat Transfer Surface, *Heat Transfer–Asian Research*, Vol. 37, No. 7, pp. 387-397.

[16] Chen, Z. and Utaka, Y., (2011). Characteristics of condensate drop movement with application of bulk surface temperature gradient in Marangoni dropwise condensation, *Int. J. Heat Mass Transfer*, Vol. 54, pp. 5049–5059.

Permissions

The contributors of this book come from diverse backgrounds, making this book a truly international effort. This book will bring forth new frontiers with its revolutionizing research information and detailed analysis of the nascent developments around the world.

We would like to thank Salim N. Kazi, for lending his expertise to make the book truly unique. He has played a crucial role in the development of this book. Without his invaluable contribution this book wouldn't have been possible. He has made vital efforts to compile up to date information on the varied aspects of this subject to make this book a valuable addition to the collection of many professionals and students.

This book was conceptualized with the vision of imparting up-to-date information and advanced data in this field. To ensure the same, a matchless editorial board was set up. Every individual on the board went through rigorous rounds of assessment to prove their worth. After which they invested a large part of their time researching and compiling the most relevant data for our readers. Conferences and sessions were held from time to time between the editorial board and the contributing authors to present the data in the most comprehensible form. The editorial team has worked tirelessly to provide valuable and valid information to help people across the globe.

Every chapter published in this book has been scrutinized by our experts. Their significance has been extensively debated. The topics covered herein carry significant findings which will fuel the growth of the discipline. They may even be implemented as practical applications or may be referred to as a beginning point for another development. Chapters in this book were first published by InTech; hereby published with permission under the Creative Commons Attribution License or equivalent.

The editorial board has been involved in producing this book since its inception. They have spent rigorous hours researching and exploring the diverse topics which have resulted in the successful publishing of this book. They have passed on their knowledge of decades through this book. To expedite this challenging task, the publisher supported the team at every step. A small team of assistant editors was also appointed to further simplify the editing procedure and attain best results for the readers.

Our editorial team has been hand-picked from every corner of the world. Their multi-ethnicity adds dynamic inputs to the discussions which result in innovative outcomes. These outcomes are then further discussed with the researchers and contributors who give their valuable feedback and opinion regarding the same. The feedback is then collaborated with the researches and they are edited in a comprehensive manner to aid the understanding of the subject.

Apart from the editorial board, the designing team has also invested a significant amount of their time in understanding the subject and creating the most relevant covers. They scrutinized every image to scout for the most suitable representation of the subject and create an appropriate cover for the book.

The publishing team has been involved in this book since its early stages. They were actively engaged in every process, be it collecting the data, connecting with the contributors or procuring relevant information. The team has been an ardent support to the editorial, designing and production team. Their endless efforts to recruit the best for this project, has resulted in the accomplishment of this book. They are a veteran in the field of academics and their pool of knowledge is as vast as their experience in printing. Their expertise and guidance has proved useful at every step. Their uncompromising quality standards have made this book an exceptional effort. Their encouragement from time to time has been an inspiration for everyone.

The publisher and the editorial board hope that this book will prove to be a valuable piece of knowledge for researchers, students, practitioners and scholars across the globe.

List of Contributors

Jan Taler
Department of Thermal Power Engineering, Cracow University of Technology, Cracow, Poland

Dawid Taler
Institute of Heat Transfer Engineering and Air Protection, Cracow University of Technology, Cracow, Poland

Mojtaba Dehghan Manshadi
Malekashtar University of Technology, Iran

Mohammad Kazemi Esfeh
University of Yazd, Yazd, Iran

Jacob O. Aweda and Michael B. Adeyemi
Department of Mechanical Engineering, University of Ilorin, Ilorin, Nigeria

Xiaohui Zhang
School of Physical Science and Technology, School of Energy, Soochow University, Suzhou, China

F.M. Hady
Department of Mathematics, Faculty of Science, Assiut University, Assiut, Egypt

S.M. Abdel-Gaied and M.R. Eid
Department of Science and Mathematics, Faculty of Education, Assiut University, The New Valley, Egypt

F.S. Ibrahim
Department of Mathematics, University College in Jamoum, Umm Al-Qura University, Makkah, Saudi Arabia

Tilak T. Chandratilleke and Nima Nadim
Department of Mechanical Engineering, Curtin University, Perth, Australia

G.A. Rivas, E.C. Garcia and M. Assato
Instituto Tecnologico de Aeronautica (ITA),Brazil

Yuzhou Chen
China Institute of Atomic Energy, China

Cheng Lin
The Second Research Division, Chung-Shan Institute of Science & Technology, Lung-Tan, Taoyan County, Taiwan, R.O.C

Yoshio Utaka and Zhihao Chen
Division of Systems Research, Faculty of Engineering, Yokohama National University, Japan

Printed in the USA
CPSIA information can be obtained
at www.ICGtesting.com
JSHW011439221024
72173JS00004B/863